INTERNATIONAL CENTRE FOR MECHANICAL SCIENCES

COURSES AND LECTURES - No. 266

ANALYSIS AND DESIGN OF ALGORITHMS IN COMBINATORIAL OPTIMIZATION

EDITED BY

G. AUSIELLO

AND

M. LUCERTINI

IASI - CNR
ISTITUTO DI AUTOMATICA
UNIVERSITA' DI ROMA

SPRINGER-VERLAG WIEN GMBH

Originally published by Springer Verlag Wien - New York in 1981

ISBN 978-3-211-81626-4 **ISBN 978-3-7091-2748-3 (eBook)**

DOI 10.1007/978-3-7091-2748-3

PREFACE

The School in Analysis and Design of Algorithms in Combinatorial Optimization was held in Udine in September 1979. It was financially supported by the National Research Council of Italy (CNR), by the International Center of Mechanical Sciences (CISM) and by the European Economic Community (CEE) and sponsored by GNASII-CNR, CSSCCA-CNR and Istituto di Automazione of the University of Rome. The organizers are pleased to express their thanks to the lecturers and participants who made the School stimulating and fruitful. Special thanks are addressed to the organizing committee and in particular to Angelo Marzollo, Alberto Marchetti-Spaccamela and Paolo Serafini who provided their friendly and valuable help making the school successful.

G. Ausiello

M. Lucertini

INTRODUCTION

The practical and theoretical relevance of problems to the NP-complete degree of complexity are widely known. From the practical point of view it is sufficient to remember that in this class we find most of the combinatorial and optimization problems which have the widest range of applications, for example scheduling problems, optimization problems on graphs, integer programming etc. As far as the theoretical relevance is concerned we should remember that one of the most outstanding problems in Computer Science, the problem of deciding whether any NP-complete set can be recognized in polynomial time, coincides with the problem of knowing whether the computation power of a nondeterministic Turing machine which accepts a set in polynomial time is strictly stronger than the power of ordinary polynomially bounded Turing machines or not. Until recently the design of algorithms for finding exact approximate solutions to practical instances of hard combinatorial and optimization problems was the main concern of experts in Operations Research while the study of the complexity of these problems with respect to various computation models and the analysis of general solution techniques was the main interest of computer scientists.

In 1978 at the Mathematisch Centrum in Amsterdam a conference was held on "Interfaces between Computer Science and Operations Research". In that occasion the experience and scientific interest of experts in mathematical programming and of Computer Scientists, interested in the design of algorithms and in the analysis of the complexity of problems, met and fruitfully interacted.

The School which was held in Udine in 1979 and which was followed by a School on "Combinatorics and Complexity of Algorithms" held in Barcellona, Spain, in 1980) is in the same framework. The program itself of the School was organized by alternating methodological results with analysis of particular classes of problems. All lectures were divided in two parts. In the first the lectures reviewed the current state of the subject and in the second they dealed with advanced topics.

More in detail Ausiello *introduced the basic concepts of machine models and computation measures with special reference to non deterministic models. Afterwards he presented a classification of many NP-complete optimization problems according to the concepts of combiatorial structure and structure preserving reductions.* Karp, *after recalling the main notions about NP-completeness theory, concentrated his attention on the probabilistical and statistical analysis of algorithms. Considering numerous examples he pointed out what are the advantages and disadvantages of his approach.* Johnson, *after reviewing the subject of approximation algorithms for NP-complete problems, lectured about strong NP-completeness and pseudopolynomiality. Then he presented the state of art of two problems: Bin packing and Graph colouring.* Paz *presented the study of NP-complete optimization problems according to a formalism introduced by himself and Moran. Following this set-up he defined rigid, simple and p-simple problems. Exploiting this classification he gave necessary and sufficient conditions for the approximability and fully approximability of NP-complete problems. Finally he defined special types of polynomial reductions showing their properties.* Sehnon *gave lectures on different subjects. Firstly he proved lower bounds in the field of algebraic complexity using algebraic geometry. Then he considered the so-called self-transformable combinatorial problems as a way to study the relationship between the search problem and the decision problem, always in the field of NP-completeness. Finally he presented some complexity properties of Boolean functions.* Luccio, *after introducing basic data structures for combinatorial problems, illustrated particular data structures and algorithms solving problems that occur in the field of multidimensional memories.* Lucertini *showed the most important algorithmic techniques to solve problems of mathematical programming. Then, he presented the group-theoretic approach to the integer programming problem.* Maffioli *introduced matroid theory and its applications to combinatorial optimization. Moreover he dealed with network design problems and, in particular, classified the complexity of various kinds of the spanning tree problem.* Lawler *presented the state of art and a generalization of network flow problems. Finally he showed Lovasz's recent algorithm for solving the linear 2-polymatroid problem.* Rinnooy Kan *lectured about advanced techniques in mathematical programming with Khachian's algorithm for solving linear programming and gave some complexity results of relevant problems of production planning. Finally* Lenstra *presented a detailed analysis of the most important results in scheduling theory and surveyed some topics in the field of routing problems.*

This volume contains papers presenting developments of some of the topics discussed by the lecturers during the School. These contributions were collected after the School with the twofold aim of giving an outlook of this important research area and of providing more detailed treatments of some specific subjects.

Giorgio Ausiello
Mario Lucertini

LIST OF CONTRIBUTORS

G. AUSIELLO	IASI-CNR, Istituto di Automatica, Università di Roma.
A. D'ATRI	Istituto di Automatica, Università di Roma.
V. FERRARI	IASI-CNR, Istituto di Automatica, Università di Roma.
M.R. GAREY	Bell Laboratories, Murray Hill, New Jersey.
S. GIULIANELLI	IASI-CNR, Istituto di Automatica, Università di Roma.
D.S. JOHNSON	Bell Laboratories, Murray Hill, New Jersey.
E.L. LAWLER	Computer Science Division, University of California, Berkeley.
J.K. LENSTRA	Mathematisch Centrum, Amsterdam.
F. LUCCIO	Università di Pisa.
M. LUCERTINI	IASI-CNR, Istituto di Automatica, Università di Roma.
F. MAFFIOLI	Istituto di Elettrotecnica ed Elettronica, Politecnico di Milano.
S. MORAN	Dept. of Mathematics, Technion, Israel Institute of Technology.
A. PAZ	Dept. of Computer Science, Technion, Israel Institute of Technology.
M. PROTASI	Istituto di Matematica, Università dell'Aquila.
A.H.G. RONNOOY KAN	Erasmus University, Rotterdam.

CONTENTS

NON-DETERMINISTIC POLYNOMIAL OPTIMIZATION PROBLEMS AND THEIR APPROXIMATION

A. PAZ* and S. MORAN**

* Dept. of Computer Science
TECHNION-Israel Institute of Technology
** Dept. of Mathematics, TECHNION-Israel Institute of Technology

1. INTRODUCTION

NP-problems are considered in this paper as recognition problems over some alphabet Σ, i.e. $A \subset \Sigma^*$ is an NP problem if there exists a NDTM (non-deterministic Turing machine) recognizing A in polynomial time. It is easy to show that the following theorem holds true.

THEOREM 1. Let A be a set in NP. Then there exists a NDTM M_A which recognizes A such that $M_A = M_{\mu_A} \circ M_{\pi_A} \circ M_{A_1}$, where

1) The operation "$\circ$" is defined as follows: $M_1 \circ M_2(x)$ is $M_1(M_2(x))$; M_1, M_2 are Turing machines and x is an input tape.

2) M_{A_1} is a polynomial time deterministic encoding machine. Its task is to encode an input $a \in A$ in some proper way to be denoted by a'.

3) M_{π_A} is a NDTM which choses some permutation $\pi(a')$ out of a possible subgroup of the group of all permutations of the encoded input tape a' in polynomial time.

4) M_{μ_A} is a polynomial time DTM which computes a number $\mu(\pi(a'))$.

5) $a \in A$ iff $\begin{cases} \mu(\pi(a')) \leq k_a & \text{(min problem)} \\ \mu(\pi(a')) \geq k_a & \text{(max problem)} \end{cases}$

where k_a is a number computed in polynomial time by the machine M_{A_1} (k_a is part of the encoding of a).

Thus every NP problem can be represented as an optimization problem and the recognition process can be split into three stages where the non-deterministic stage (the machine M_{π_A}) is separated from the other stages.

2. NP OPTIMIZATION PROBLEMS

The conjecture that $P \neq NP$ is widely believed to be true. This conjecture prompted many researchers to develop and study polynomial approximations for problems in NP, when considered as optimization problems. See e.g. [Jo 73] or [Sa 76].

The previous section points toward the possibility of a new approach to the study of NP problems and NP optimization problems. In what follows, an attempt is made to develop that new approach. The results achieved so far are promising. These results provide some new insight into recently proved approximation results and it is hoped that they will serve as a basis for a more extensive theory of combinatorial approximations.

DEFINITION 1. An NP optimization problem (N POP) is a subscripted 4-tuple $(A,F,t,\mu)_{EXT}$ where:

$$EXT = MIN \text{ or } EXT = MAX.$$

$A \subset \Sigma^*$ is a polynomial time recognizable recursive set over a finite alphabet Σ (A is the set of all well formed encodings of some given combinatorial entity e.g. graph, family of sets, logical sentence in CNF, etc.). It is assumed that $\lambda \in A$ where λ denotes the empty word.

F is a function $F:A \to P_0(A)$ (the set of all finite subsets of A), where for all $a \in A, F(a)$ is a subgroup of the group of all permutations of a, to be called "the set of proper permutation of a". An element in F(a) will be denoted by $\pi(a)$. It is also assumed that the many valued function $a \to F(a)$ is computed in polynomial time by a NDTM ("*permutation machine*").

t is a function $t:A \to P_0(Z \cup \{\pm\infty\})$.t is a function intended to specify the property of the elements of A we want to study e.g. the number of clauses which are satisfiable in a given CNF formula, the number of nodes that are pairwise adjacent in a given graph, etc. With regard to the function t we shall use the following notation

$$op(a) = \text{optimum}(k:k \in t(a)) \text{ where}$$

optimum is "max" if EXT = MAX and it is "min" if EXT = MIN. We shall use the value $-\infty$ in connection with MAX problems and the value $+\infty$ in connection with MIN problems.

It is also assumed that F is compatible with t, that is: $a' \in F(a)$ implies that $t(a') = t(a)$, that $t(\lambda) = \{0\}$, and that $t(a) \neq \emptyset$ for all $a \in A$.

μ is a polynomial time function (the measure function) $\mu : \Sigma^* \to Z \cup \{\pm\infty\} \cup \{\alpha\}$ $(\alpha \notin Z)$ satisfying the following properties:

$$\mu(w) = \alpha \qquad \text{iff} \quad w \notin A; \tag{1}$$

$$(\mu(a) = k) \to k \in t(a); \tag{2}$$

$$(\forall a \in A)(\exists \pi^*(a) \in F(a)) \quad (\mu(\pi^*(a)) = op(a)). \tag{3}$$

It should be noticed that the combinatorial properties (and the complexity) of a given NPOP are determined by A, t and the subscript EXT. We shall therefore abbreviate our notation and use the notation $(A,t)_{EXT}$ or $(A,t,\mu)_{EXT}$ whenever the other parameters are not relevant to the context.

Examples: (1) The problem mentioned before MAX SAT can be described in the form $(A,t)_{MAX}$ where A is the set of all CNF formulas and for $a \in A$, $k \in t(a)$ iff there is a truth assignment to the variables occuring in a which satisfies exactly k clauses.

(2) Colorability: $(\mathcal{G},t)_{MIN}$ where $\mathcal{G}$ is the set of all graphs and for $G \in \mathcal{G}$, $k \in t(G)$ iff G is k-colorable.

(3) Let A be any set in NP. One can show that A can be defined by the NPOP $(\Sigma^*, t_A)_{MAX}$, where $t_A(w) \subset \{0,1\}$ for all $w \in \Sigma^*$, and $1 \in t_A(a) \Leftrightarrow a \in A$. It follows that $A=\{w \mid op(w)=1\}$.

REMARK. When considering NP problems as recognition problems a distinction should be made between "*polynomially constructive*" solutions and "*polynomially nonconstructive*" solutions.

Considering e.g. colorability: it is clear that the problem

of ascertaining whether a given graph is k-colorable is different from the problem of actually finding a k coloration (provided that it exists). This suggests the necessity of distinguishing "*constructive*" and "*nonconstructive*" solutions of NPOP's.

Let $(A,F,t,\mu)_{EXT}$ be a NPOP. An "*algorithm that solves* $(A,t)_{EXT}$" is a recursive function $f: \Sigma^* \to Z \cup \{\pm\infty\} \cup \{\alpha\}$ $(\alpha \notin Z)$ satisfying the following:

1) $f(w) = \alpha \leftrightarrow w \notin A$

2) $(\forall a \in A)\ (f(a) = op(a))$.

An "*algorithm that solves,* $(A,t)_{EXT}$ *constructively*" is a recursive function $f: \Sigma^* \to \Sigma^*$ satisfying the following:

1) $f(w) = \beta \leftrightarrow w \notin A$ (β is a string not in A)

2) $(\forall a \in A)\ (f(a) = \pi^*(a))$.

DEFINITION 2. $(A,t)_{EXT}$ is "*(constructively) polynomially solvable*" if there exists a polynomial time algorithm that solves $(A,t)_{EXT}$ (constructively), and such an algorithm is a "*(constructive) polynomial solution*" of $(A,t)_{EXT}$.
We shall show now that the two notions of solvability are equivalent to each other in some global sense.

LEMMA 2. $(A,F,t,\mu)_{EXT}$ is constructively polynomial solvable implies that $(A,t)_{EXT}$ is polynomially solvable.

PROOF. Let f be a constructive polynomial solution of $(A,t)_{EXT}$. Define f' by:

$$f'(w) = \begin{cases} \alpha & w \notin A \\ \mu(f(w)) & w \in A \end{cases}$$

It follows directly from the definitions that f' is a polynomial algorithm that solves $(A,t)_{EXT}$.

THEOREM 3. (a) If all NPOP's are polynomially solvable then P = NP.

(b) If P = NP then all NPOP's are constructively polynomially solvable.

PROOF. Part (a) follows easily from Theorem 1. Part (b) will be proved in 2 stages. We shall show first that if P = NP then all NPOP's are polynomially solvable, and then we shall show that for each $NPOP(A,t)_{EXT}$ there exists an $NPOP(A,t')_{EXT}$ such that a polynomial solution of the second provides a constructive polynomial solution of the first.

To prove the first part, we note that a polynomial time measure machine can compute, on input a, integers that are not larger than $2^{P(1(a))}$(*) for some polynomial P. This implies that $op(a) \leq 2^{P(1(a))}$. Moreover, for each $k \leq 2^{P(1(a))}$, the set $\{a \in A | op(a) \leq k\}$ can be recognized by a nondeterministic polynomial time Turing machine. (This follows directly from the definition of NPOP). Therefore, using binary search, no more than P(1(a)) NP recognition problems of the form "is $op(a) \leq k$?" have to be solved in polynomial time, and therefore op(a) can be found in polynomial time.

To prove the second part, we assume W.L.G. that $\Sigma = \{0,1\}$. Let n(a) be the binary number represented by a string a $(a \in \Sigma^+)$, e.g. n(0010) = 10. Let $(A,F,t,\mu)_{EXT}$ be a given NPOP and assume the output of the measure function μ to be given in binary numbers. Define a new measure function μ' as follows: $\mu'(a) = 2^{1(a)}\mu(a)+n(a)$ (e.g. if a = 0010 and $\mu(a) =$ = 101 then $\mu'(a) = 1010010$). Finally let t'(a) be defined as $t'(a) = \{k | (a' \in F(a))(\mu'(a') = k)\}$. One verifies easily that:

(*) 1(a) indicates the lenght of the string a.

1) $\mu(a) < \mu(a') \rightarrow \mu(a) < \mu'(a')$;

2) If $\pi^*(a)$ is an optimal permutation for $(A,t')_{EXT}$ then it is also an optimal permutation for $(A,t)_{EXT}$.

3) If $(A,t)_{EXT}$ is an NPOP then so is $(A,t')_{EXT}$.

4) For any $a \in A$, op'(a) (=op(a) according to t') is a number such that its last 1(a) digits represent $\pi^*(a)$, and the other digits are (the binary representation of) op(a).

From the above discussion it is evident that any solution for $(A,t')_{EXT}$ provides a constructive solution for $(A,t)_{EXT}$, since knowing op'(a) is equivalent to the knowing of $\pi^*(a)$ and op(a) at the same time

QED

Combining Lemma 2 with Theorem 3 we have the following:

COROLLARY 1. The following three conditions are equivalent:

(1) P = NP

(2) All NPOP are polynomially solvable.

(3) All NPOP are constructively polynomially solvable.

PROOF. By Theorem 3 (2) implies (1) implies (3). By Lemma 2 (3) implies 2.

QED

Another interesting consequence of Lemma 2 and Theorem 3 is reflected in the following definition and corollary:

DEFINITION 3. An NPOP is NPOP complete if the existence of a polynomial solution to it implies that P = NP.

COROLLARY 2. If some NPOP complete problem is polynomially solvable then all NPOP's are construvtively polynomially solvable, in addition a given NPOP complete problem is polynomially solvable if and only if it is constructively polynomially solvable.

REMARK 1. It is clear that every NP complete problem when viewed as an NPOP, and every NPOP problem such that a polynomial solution to it would imply that some NP complete problem is polynomially solvable, are NPOP complete.

REMARK 2. Corollary 2 justifies Definition 3 which is shown, in this corollary, to be in accordance with the definition of NP-complete problems adopted by Aho-Hopcrofts Ulman [AHU 74, p.373].

3. REDUCTIONS BETWEEN NPOP's

On the basis of the previous definitions we are able to define and study reducibility and in particular polynomial reducibility between NPOP's.

DEFINITION: Let $(A_1,t_1)_{EXT_1}$ and $(A_2,t_2)_{EXT_2}$ be two NPOP's. Then g: $\Sigma^* \to \Sigma^*$ is a (polynomial) reduction of the first NPOP into the second iff g (is a polynomial function which) satisfies the following conditions:

(1) $a_1 \in A_1$ iff $g(a_1) \in A_2$;

(2) There exists a (polynomial time) function $\delta: A_1 \times Z \to Z$ such that: $\forall a_1 \in A_1$, $\delta(a_1, op(g(a_1))) = op(a_1)$ (that is, one can compute $op(a_1)$ if $op(g(a_1))$ is known). For abbreviation we shall use the following notation: given $(A,E,t,\mu)_{EXT}$, then for $a \in A$; $k_1, k_2 \in t(a)$:

$$k_1 \overset{*}{<} k_2 \;\text{<->}\; \begin{cases} k_1 < k_2 & \text{EXT = MIN} \\ k_1 > k_2 & \text{EXT = MAX} \end{cases}$$

(that is if k_1 is a better approximation to op(a) than k_2). The reduction is *order preserving* if the above function satisfies the following additional conditions:

Let $a_1 \in A_1$ and $a_2 = g(a_1) \in A_2$, then

$$\forall k \in t_2(a_2),\ \delta(a_1,k) \in t_1(a_1)\ ; \tag{3.1}$$

$$\forall k_1,k_2 \in t_2(a_2) \text{ it is true that} \tag{3.2}$$

$$k_1 \overset{*}{<} k_2 \;\text{<->}\; \delta(a_1,k_1) \overset{*}{<} \delta(a_1,k_2).$$

An example of order preserving reduction is given by ([Ka 72]): Let MAX CLIQUE be the following NPOP:

$(G,t_{MC})_{MAX}$ where G is the set of all graphs, and for $G \in G$, $t_{MC}(G) = \{k|G$ contains a complete subgraph with k nodes$\}$.

Let NODE COVER be the following $(G,t_{NC})_{MIN}$ problem: G is the set of all graphs, as before, and for $G \in G, t_{NC}(G) = \{k$ |there exist k nodes in G which are incident to all arcs of G$\}$.

The following reduction g: $G(N,A) \rightarrow G'(N,\bar{A})$ where $\bar{A} = \{(i,j)|(i,j) \notin A\}$ is an order preserving reductions of MAX CLIQUE to NODE COVER and vice versa, with the following δ: $\delta(G(N,A),k) = |N| - k$.

An order preserving reduction is *measure preserving* if the function δ satisfies also the property:

$$(\forall a_1 \in A_1)(\forall k \in Z),\ \delta(a_1,k) = k. \tag{3.3}$$

The measure preserving reductions have the property that any measure μ_2 on $(A_2,t_2)_{EXT_2}$ induces a measure μ_1 on $(A_1,t_1)_{EXT_1}$ such that $\mu_1(a_1) = \mu_2(a_2)$ ($a_1 \in A_1$ and $a_2 = g(a_1) \in A_2$). It is easy to show that measure preserving reductions can exist only between NPOPs such that $EXT_1 = EXT_2$. The importance of measure preserving reductions will be illustrated in Section 4, Lemma 6.

The notation "$(A,t_1)_{EXT_1} \underset{-}{\leq} (B,t_2)_{EXT_2}$" will be used to denote measure preserving reducibility, where "$\underset{p}{\leq}$" denotes polynomial reducibility and "$\overset{g}{\underset{p}{\leq}}$" denotes polynomial reducibility with corresponding function g. The relation $\underset{p}{\leq}$ is reflexive and transitive.

Constructive reductions

DEFINITION 4. For any given NPOP $(A,F,T,\mu)_{EXT}$ for all $a \in A$ and every pair $a_1, a_2 \in F(a)$

$$a_1 \overset{**}{<} a_2 \Leftrightarrow \mu(a_1) \overset{*}{<} \mu(a_2).$$

Thus $\pi^*(a) \overset{**}{\leq} \pi(a) \in F(a)$ for all $\pi(a) \in F(a)$. The relation $\overset{**}{\leq}$ induces a partial order on F(a) for all $a \in A$.

In many cases we would like to find for a given NPOP and $a \in A$, a $\pi(a)$ such that $\pi(a) \overset{**}{<} a$ rather than having only a value k having the property that $k \overset{*}{<} \mu(a)$. Thus if the NPOP is a MAX CLIQUE we would like to *get* as big a clique as possible in a given graph rather than ascertaining that the graph *has* a clique containing k nodes. This motivates the following:

DEFINITION 4. Let $(A_1,F_1,t_1,\mu_1)_{EXT_1}$ and $(A_2,F_2,t_2,\mu_2)_{EXT_2}$ be two NPOP's. $g: \Sigma^* \to \Sigma^*$ is a constructive (polynomial) reduction of the first NPOP to the second iff g is a (polynomial) function satisfying condition (1) of Definition 4, together with the following condition:

(2') There exists a (polynomial time) function $f^C: A_1 \times A_2 \to A_1$ such that:
$(\forall a_1 \in A_1)\ f^C(a_1, \pi^*(g(a_1))) = \pi^*(a_1)$ (that is: given a_1, one can compute $\pi^*(a_1)$ if $\pi^*(g(a_1))$ is known). A constructive reduction is *order preserving* if the above function f^C satisfies also the additional conditions:

Let $a_1 \in A_1$ and $\mu_2 = g(a_1) \in A_2$, then:

$$\forall a_2' \in F_2(a_2),\ f^C(a_1, a_2) \in F_1(a_1). \tag{4.1}$$

$$(\forall a_2', a_2'' \in F_2(a_2))\ a_2' \overset{**}{<} a_2'' \to f^C(a_2', a_1) \overset{**}{<} f^C(a_2'', a_1) \tag{4.2}$$

A constructive order preserving reduction is *measure preserving* if the function f^C satisfies the following:

$$(\forall a_1 \in A_1)(\forall a_2' \in F(a_2))\ \mu_2(a_2') = \mu_1(f^C(a_1, a_2')). \tag{4.3}$$

It can be shown that the existence of constructive reductions of one of the three types above implies the existence of nonconstructive reductions of the same type, the proof being similar to the proof of Lemma 2

Some Examples. We give now, without proof, some measure preserving reduction between NPOP's. All of the following reductions can be shown to be constructive reductions. Part of the reductions are Karp's reductions ([Ka 72]) when adjusted to NPOP's. For definitions of the NPOP's see appendix.

(i) g_1 : COLORABILITY → CLIQUE COVER: a graph G(N,A) is reduced to the complemented graph $G'(N,\bar{A})$.

(ii) g_1^{-1} : CLIQUE COVER → COLOROBILITY, is also a measure preserving reduction.

(iii) g_2 : SET COVER → DOMINATING SET: An input to SET COVER of the form $\phi = \{S_1 \ldots S_n\}$, where $\bigcup_{i=1}^{n} S_i = S = \{x_1,\ldots,x_m\}$ is reduced to a graph G(N,A), where:

$$N = \{1,2,\ldots,n,x_1,x_2,\ldots,x_m\}$$

$$A = \{(i,j)\,|\,1 \leq i < j \leq n\} \cup \{(i,x_t)\,|\,x_t \in S_i\}.$$

(iv) g_2' : DOMINATING SET → SET COVER: An input to DOMINATING SET of the form G(N,A) is reduced to a family of sets ϕ in the following manner: Suppose $N = \{1,2,\ldots,n\}$ then $\phi = \{S_1,S_2,\ldots,S_n\}$ where $S_i = \{i\} \cup \{j\,|\,(i,j) \in A\}$.

(v) g_3 : NODE COVER → DOMINATING SET: An input G(N,A) to NODE COVER is transformed to G'(N',A') where:

$$N' = N \cup A$$

$$A' = \{(i,j)\,|\,i,j \in N\} \cup \{(i,e)\,|\,i \in N,\ e \in A,\ i \text{ incident to } e\}.$$

(vi) g_4 : MAX SAT → MAX CLIQUE: An input to MAX SAT of the form $\{C_1,\ldots,C_p\}$, where each C_i is a clause over a set of variables $\{X_1,\bar{X}_1,\ldots,X_n,\bar{X}_n\}$ is reduced to a graph G(N,A), where:

$$N = \{V_{\sigma i}\ |\ \sigma \text{ is a literal, } \sigma \in C_i\}$$

$$A = \{(V_{\sigma i},V_{tj})\ |\ t \neq \bar{\sigma},\ i \neq j\}.$$

(vii) g_5 : NODE COVER → SET COVER: An input to NODE COVER of the form G(N,A) is reduced to $\phi = \{S_i\}_{i \in N}$, where $S_i = \{i,j)\,|\,(i,j) \in A\}$ (note that the existence of g_5 follows from the existence of g_3 and g_2').

(viii) g_6 : NODE COVER → FEEDBACK NODE SET: A graph $G(N,A)$ is reduced to a digraph $D(V,E)$ where

$$V = N$$

$$E = \{(i \to j),(j \to i) \in A\}.$$

(ix) g_1 : NODE COVER → FEEDBACK ARC SET: A graph $G(N,A)$ is reduced to a digraph $D(V,E)$ where $V = \bigcup_{i \in N} \{i_1, i_2\}$

$$E = \bigcup_{(i,j) \in E} \{(i_1 \to i_2),(i_2 \to j_1)(j_1 \to j_2),(j_2 \to i_1)\}$$

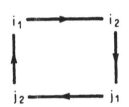

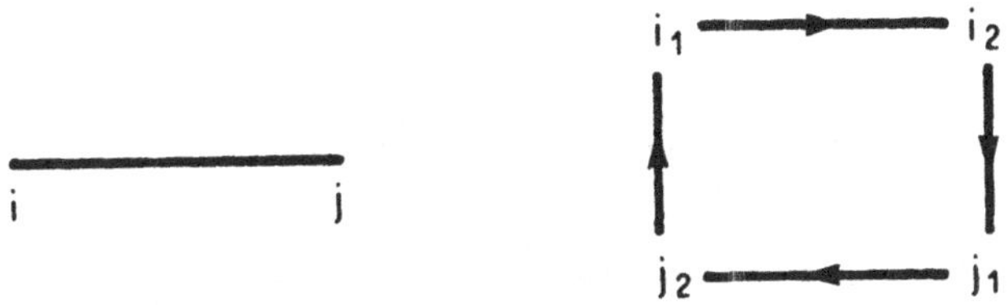

The following diagram illustrates the above reductions:

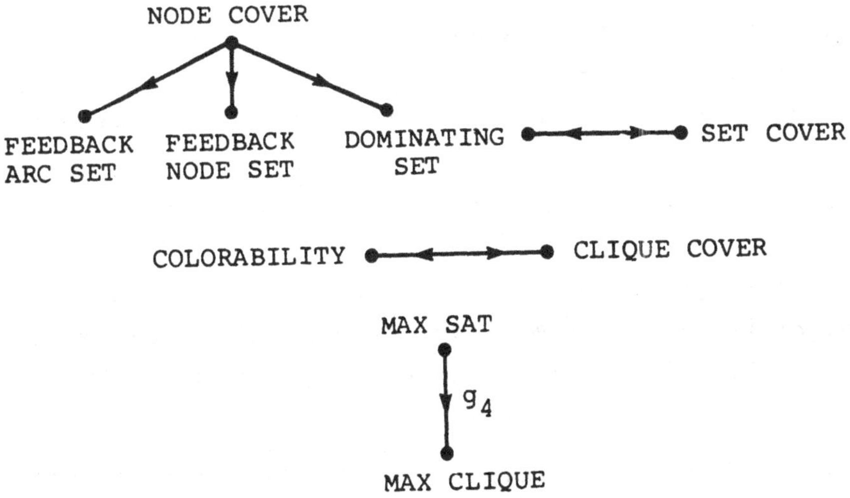

It will be shown now that the class of NPOP's can be divided into two subclasses, such that no problem in one class can be reduced by a measure preserving reduction to a problem in the second class (unless P = NP).

DEFINITION 5. Let $(A,t)_{EXT}$ be a NPOP. Then for each $k \in Z$, $(A,t)_{EXT,k} = \{a | a \in A \text{ and } op(a) \leq k\}$.

DEFINITION 6. $(A,t)_{EXT}$ is a "*simple NPOP*" iff for all $k \in Z$, $(A,t)_{EXT,k}$ is a set in P. It is a "*rigid NPOP*" if it is not simple (i.e. for some k, $(A,t)_{EXT,k}$ is in NP\P, where the notation NP\P stands for the sets which are in NP and are not in P provided that P ≠ NP).

THEOREM 4. If $(A,t_1)_{EXT}$ is a rigid NPOP and $(B,t_2)_{EXT}$ is a simple NPOP, then

$$(A,t_1)_{EXT} \underset{p}{\not\leq} (B,t_2)_{EXT}.$$

PROOF. Let $k_0 \in Z$ be such that $(A,t_1)_{EXT,k_0} \in NP\backslash P$. Assume that $(A,t_1)_{EXT} \overset{g}{\underset{p}{\leq}} (B,t_2)_{EXT}$. The following polynomial algorithm will check for each $w \in \Sigma^*$ if $w \in (A,t_1)_{EXT,k_0}$:

(a) check if $w \in A$, if not reject;

(b) reduce w by g to $b \in B$;

(c) check whether $b \in (B,t_2)_{EXT,k_0}$. If so accept else reject.

(Clearly, $b \in (B,t_2)_{EXT,k_0} \iff w \in (A,t_1)_{EXT,k_0}$).

All three steps of the algorithm are polynomial, so that the algorithm is polynomial as a whole. It follows that $(A,t_1)_{EXT,k_0} \in P$, which is impossible. The theorem is thus proved.

REMARK. Theorem 4 will remain true if the definition of

measure preserving reductions is generalized as follows:

Replace condition 3.3 by the condition

3.3' $(\forall a_1 \in A_1)\ (\forall k \in Z)(\mathfrak{f}(a_1,k) = \eta(k))$

where η is a (polynomial time) function from Z to Z.

Similar generalizations are possible for the other theorems concerning measure preserving reductions given in the sequel.

If $P \neq NP$, then a set A is NP complete implies $A \in NP\backslash P$. Combining this with the known NP completeness results, all known NPOP's can be shown to be either rigid or simple. Some examples are given below:

RIGID NPOP's

(a) Colorability (see [S 73]). *Planar* colorability is a special type of rigid NPOP, as there is only one k(=3), for which $(A,t)_{EXT,k}$ is NP complete.

(b) Bin Packing = $(IS,t_{BP})_{MIN}$

$$IS = \{(a_1,\ldots,a_n,a_{n+1}) \mid \forall i\ a_i \in Z\}$$

$t_{BP}((a_1,\ldots,a_n,a_{n+1})) = \{k \mid$ the set $\{a_1,\ldots,a_n\}$ can be divided into k subsets, the sum of numbers in each of them $\leq a_{n+1}\}$.

SIMPLE NPOP's

(a) MAX SAT
(b) MAX CLIQUE
(c) SET COVER
(d) DOMINATING SET
(e) NODE COVER
(f) FEEDBACK NODE SET
(g) FEEDBACK ARC SET
(h) MAX SUBSET SUM.

A NPOP complete by transformation problem

In the note of Knuth [Kn 74] a distinction is made between NP complete and NP complete by transformation problems, where the former is a set such that if it is in P then P=NP, and the later is a set such that all NP sets can be reduced to it by a polynomial time reduction.

A similar distinction can be made for NPOP's where a NPOP complete by transformation problem is a problem such that all NPOP's can be reduced to it by a measure preserving (polynomial) reduction. (This kind of reduction is chosen due to its properties with regard to approximation algorithm - see Lemma 6). For the NPOP case it follows from the preceding sections that if $P \neq NP$ then the NPOP complete by transformation set is properly included in the NPOP complete set. Such a (proper) inclusion has not been yet proved (or disproved) for the NP case.

Following the same note of Knuth, we define an "*NPOP hard problem*" as an optimization problem such that a polynomial solution to it would imply that P = NP. A NPOP hard problem does not have to be a NPOP.

We present here a set A and corresponding function t such that both $(A,t)_{MIN}$ and $(A,t)_{MAX}$ are NPOP's and, for each NPOP $(B,t)_{EXT}$, $(B,t)_{EXT} \underset{p}{\leq} (A,t)_{EXT}$ where EXT = MAX or EXT = MIN. Such an NPOP is NPOP complete by transformation problem. Our example will, therefore, provide an analogue to Cook's theorem (for recognition problems) for NPOP's. As a matter of fact, the example we are going to present is an extension of the example of Cook made to fit our definitions. We first restate Cook's theorem (without proof) in a slightly different form suitable for our purpose.

Theorem of Cook

Let T be an NDTM, and let $f: Z \to Z$ be a (polynomial time) function, $f(n) \geq n$. Then there exists a function $g: \Sigma^* \to \Sigma^*$ that satisfies the following conditions:

(1) $g(w) \in SAT \leftrightarrow w$ is accepted by T within $f(1(w))$ steps.

(2) The time complexity of g is $p(f(1(w)))$ for some fixed polynomial $p(n)$ ($p(n) < 0(n^4)$). (In the original theorem of Cook f is the polynomial representing the time complexity of T).

Let $(W(CNF), t_{COM})$ be a set and corresponding t-function where: W(CNF) is the set of all logical formulas in Conjuctive Normal Form over some set of variables X, combined with a weight function $W: X \to Z$.

For a given $a \in W(CNF)$, we define $t_{COM}(a)$ as follows:

Let $B_a = \{B | B: X_a \to \{0,1\}$ is a valuation of the set X_a of the variables appearing in a$\}$ $(B(\sigma) = 1 \leftrightarrow B(\bar{\sigma}) = 0)$.

Define a function $M_{COM}: B_a \to Z \cup \{\pm\infty\}$ as

$$(\forall B \in B_a) M_{COM}(B) = \begin{cases} \pm\infty & \text{if B does not satisfy the logical formula a} \\ \sum_{x \in X_a} W(x)B(x) & \text{else.} \end{cases}$$

then: $$t_{COM}(a) = \bigcup_{B \in B_a} \{M_{COM}(B)\}.$$

DEFINITION 7. An "*NP measure function*" is a function $\mu: \Sigma^* \to P_0[Z \cup \{\pm\infty\} \cup \alpha]$ $(\alpha \notin Z)$ that can be computed by a nondeterministic polynomial time Turing machine. ("NP measure" machine). Let $(A,F,t,\mu)_{EXT}$ be a NPOP. By Definition 1 there exists an NP measure machine, T, such that for each $a \in A$, $k \in \mu(F(a)) \leftrightarrow$ there exists a legal computation of T which terminates within P(1(a)) steps, in an accepting state, with k written on its tape. Moreover, we may assume that k is printed in binary digits in reverse order, i.e. if k = $= \sigma_1\sigma_2 \dots \sigma_r$, $\sigma_i = 0$ or $\sigma_i = 1$, then the output on the tape will be $\sigma_r\sigma_{r-1} \dots \sigma_1$.

THEOREM 5. (Cook theorem for NPOP's): Let T be a non-deterministic measure machine, and let $f: Z \to Z$ be a recursive (polynomial time) function ($f(n) \geq n$). Then there exists a recursive function $g: \Sigma^* \to \Sigma^*$ such that:

(1) $g(w) \in W(CNF)$ and $k \in t_{COM}(g(w))$ <-> there exists a legal computation of T which terminate within $f(1(w))$ steps in an accepting state with k written on its tape.

(2) The time complexity of g is $p(f(1(w)))$, where $p(n) < 0(n^4)$ is some fixed polynomial.

PROOF. WLG we may assume that T has the properties described above (i.e. prints the output in reverse order). For a given $w \in \Sigma^*$, we define the reduction g as follows:

(1) Perform the usual reduction of Cook for w. As a result one gets a logical formula in CNF, over some set of variables X.

(2) Define a weight function on X in the following way: Let $C(i,1,f(1(w)))$ be the variables in Cook's reduction which *asserts* that the symbol 1 is written in cell i at time $f(1(w))$, $i = 0,1,\dots,f(1(w))$. Now, for all $x \in X$:

$$W(x) = \begin{cases} 2^i & \text{if } x = C(i,1,f(1(w))) \\ 0 & \text{else} \end{cases}$$

We now claim that:

(1) for $a \in \Sigma^*$, $g(a) \in SAT$ <-> on input a T halts in an accepting state within $\delta(1(w))$ steps. (This is, in fact, Cook's Theorem).

(2) For $a \in \Sigma^*$, $k \in M_{COM}(g(a))$ <-> on input a there exists a legal computation of T which terminates in an accepting state within time $\delta(1(a))$, with k written on its tape, in reverse order, in binary digits. (2) follows from (1) and from the definition of the weight function W.

QED

REMARK 1. The time required for the above reduction differs from that of Cook's original reduction by at most $O(\delta(1(a))^2)$ steps required to define the weight function W.

REMARK 2. It can be shown that every NPOP can be reduced by a *constructive* measure preserving reduction to (W(CNF), t_{COM}) by requiring that the NP measure machine will write not only op(a) but also $\pi^*(a)$ at the end of the computation. (The details are similar to those of Theorem 2). The reader is referred to [HB 76] for related topics.

Other properties of reductions between NPOP's will be discussed in a forthcoming paper.

4. P-APPROXIMATION FOR NPOP

The last section of the paper will deal with the problem of approximating NPOP's in polynomial time.

DEFINITION 8. A function $h: \Sigma^* \to Z \cup \{\pm\infty\} \cup \{\alpha\}$ is a p-approximation for an NPOP $(A,t)_{EXT}$ iff h is a polynomial time function satisfying the following properties:

(1) $h(w) = \alpha$ iff $w \notin A$;

(2) $h(a) \geq op(a)$ if EXT=MIN and $h(a) \leq op(a)$ if EXT = MAX.

DEFINITION 8. A function $h^c: \Sigma^* \to \Sigma^*$ is a constructive p-approximation for an NPOP$(A,F,t,\mu)_{EXT}$ iff h^c is a polynomial (in the length of a) time function satisfying the following properties:

$h^c(w) = \alpha'$ iff $w \notin A$ (' is a string not in A);

$(\forall a \in A) h^c(a) \in F(a)$.

In what follows, we shall state the results for both constructive and non-constructive cases, but the proof for the constructive cases, in general.

The performance of a p-approximation h can be defined as follows (see [Sa 76]:

$$(\forall a,\ a \neq P_{h,(A,t)_{EXT}}(a) = \left\{ \frac{|h(a)-op(a)|}{\min(h(a),op(a))} \right\}$$

And as a function of the length of the input the performance is defined as:

$$(\forall n \in Z) p_{h(a,t)_{EXT}}(n) = \max\{P_{h(A,t)_{EXT}}(a) \mid \ell(a) \leq n\}. \quad (1(a)=\text{length of } a).$$

DEFINITION 9. An NPOP$(A,t)_{EXT}$ is $\ell(n)$ p-approximable if there is a p-approximation function h for $(A,t)_{EXT}$ such that $P_{h,(A,t)_{EXT}}(n) \leq \ell(n)$ for all $n \in Z$.

An NPOP $(A,t)_{EXT}$ is p-approximable iff for any $\varepsilon > 0$ there is a p-approximation function h for $(A,t)_{EXT}$ such that $P_{h(A,t)}(a) \leq \varepsilon$ for all $a \in A$ $(A,t)_{EXT}$ is fully p-approximable iff for any $\varepsilon > 0$ there is a p-approximating function h as above with the additional property that h can be computed in polynomial time Q where $Q = Q(\ell(a), \frac{1}{\varepsilon})$ i.e. Q is polynomial in both the length of a and the value $\frac{1}{\varepsilon}$. $(A,t)_{EXT}$ is constructive p-approximation.

The importance of measure preserving reductions follows from the following:

Lemma 6. If $(A,F_1,t_1,\mu_1)_{EXT} \underset{p}{\overset{g}{\leq}} (B,F_2,t_2,\mu_2)_{EXT}$ then the following holds true; If $(B,t_2)_{EXT}$ is (fully) p-approximable then so is $(A,t_1)_{EXT}$. If g is a constructive reduction and $(B,t_2)_{EXT}$ is constructively (fully) p-approximable then so is $(A,t_1)_{EXT}$.

PROOF. Let the time complexity of g be $P_0(n)$, for some polynomial P_0. Then, by definition, for all $a \in A, 1(g(a)) <$ $\leq P_0(1(a))$.

Assume that (B,t_2) is fully p-approximable in $P(1(a),\frac{1}{\varepsilon})$ time for some polynomial P. One can assume that P is non-decreasing in both its variables, theorwise the negative terms may be omitted. We must show that $(A,t_1)_{EXT}$ is fully p-approximable in $P'(A(a),\frac{1}{\varepsilon})$ time for some (other) polynomial P'.

Let $a \in A$ and $\varepsilon > 0$ be given. Then we can find an ε approximation to op(a) using the following algorithm:

(1a) Reduce a by g to $b \in B$;

(1b) Find an ε-approximation to op(b)=op(a) (g is measure preserving).

Due to the fact that for measure preserving reductions $EXT_1 = EXT_2$, every ε-approximation to op(b) is also an ε-approximation to op(a) (both approximations will have the same value and op(a)=op(b)). The time required by Step (1a) of this algorithm is bounded by $P_0(1(a))$ and the time required by Step (1b) of the algorithm is bounded therefore by $P(1(b),\frac{1}{\varepsilon}) \leq P(P_0(1(a)),\frac{1}{\varepsilon}) = P_1'(1(a),\frac{1}{\varepsilon})$ (P was assumed to be nondecreasing) $P'(1(a), \frac{1}{\varepsilon})$ is the required polynomial.

REMARK: This proof will fit also the p-approximation case with some minor changes which are left to the reader.

The "constructive" part of the lemma follows from the fact that, if $(B,t_2)_{EXT}$ is constructively fully p-approximable, then (1b) of the algorithm can be changed (1b'): Find $b' \in F_2(b)$ such that $\mu_2(b')$ is an ε approximation to op(b).

g being a constructive reduction, one can find (in a polynomial time) an $a' \in F_1(a)$ such that $\mu_1(a') = \mu_2(b')$ and since op(a) = op(b), $\mu_1(a')$ is an ε-approximation to op(a).

REMARK 1: See remark after Theorem 4.

REMARK 2: Lemma 6 can be extended in two ways:

(a) if $(B,t_2)_{EXT}$ is C p-approximable for some constant C then so is $(A,t_1)_{EXT}$.

(b) Let $\delta(n)$ be a function that satisfies the following: For each $k \in Z$, there exists a constant C_k such that for all $n \in Z$, $\frac{\delta(n^k)}{\delta(n)} \leq C_k$ (e.g. $\delta(n) = (\log n)^r$ is such a function with $C_k = k^r$.

Then if $(B,t_2)_{EXT}$ is $\delta(n)$ p-approximable, then $(A,t_1)_{EXT}$ is $O(\delta))$ p-approximable. The proof is omitted and this paper will deal only with p-approximation and fully p-approximation.

Some results concerning p-approximable, and in particular p-approximable NPOP's, are represented below.

4.1. *Necessary Condition for p-Approximability*

THEOREM 7. If $(A,t)_{EXT}$ is p-approximable then $(A,t)_{EXT}$ is simple.

PROOF. Let $(A,t)_{EXT}$ be p-approximable, and let $k \in Z$ be given. Then $(A,t)_{EXT,k}$ is in P: by definition, for each $\varepsilon \geq 0$ there is a polynomial (time) function $h: \Sigma^* \to \Sigma \cup \{\pm\infty\} \cup \{\alpha\}$

such that $\forall a \in A$, $\frac{|h_\varepsilon(a) - op(a)|}{\min\{h_\varepsilon(a), op(a)\}} < \varepsilon$.

Let EXT = MAX. (The other case is similar and is omitted). $h_\varepsilon(a)$ and $op(a)$ are integers by definition and $h(a) \leq op(a)$. Thus, $h_\varepsilon(a) > k$ implies that $op(a) > k$ on the other hand, choosing $\varepsilon = \frac{1}{k}$, the inequality

$$\frac{|h_\varepsilon(a) - op(a)|}{\min\{h_\varepsilon(a), op(a)\}} < \frac{1}{k} \text{ implies that } \frac{op(a) - h_\varepsilon(a)}{h_\varepsilon(a)} < \frac{1}{k} \text{ or } \frac{op(a)}{h_\varepsilon(a)} - 1 < \frac{1}{k}$$

and for $h_\varepsilon(a) \leq k$ this inequality is impossible unless $op(a) = h(a)$. It follows that:

$$[h_{\frac{1}{k}}(a) \leq k] \longleftrightarrow [op(a) \leq k].$$

In other words, $h_{\frac{1}{k}}$ is polynomial function that recognizes $(A,t)_{EXT,k}$.

QED

It can be shown that the converse of Theorem 7 is not true, and that there are some simple NPOP's which are not p-approximable (the TSP(*) problem [PS 76] is an example), assuming $P \neq NP$.

4.2. *Necessary Condition for Fully p-Approximability*

DEFINITION 10. $(A,t)_{EXT}$ is p-simple iff there is some

(*) See Appendix.

polynomial $Q(x,y)$ such that $\forall k \in Z$, $(A,t)_{EXT,k}$ is recognizable in $Q(1(a),k)$ time.

THEOREM 8. $(A,t)_{EXT}$ is fully p-approximable implies that $(A,t)_{EXT}$ is p-simple.

PROOF. Let $(A,t)_{EXT}$ be fully p-approximable and let $k \in Z$ be given. Then $(A,t)_{EXT,k}$ is recognizable in $Q(1(a),k)$ time for some polynomial $Q(x,y)$: by definition there is some polynomial $Q'(x,y)$ such that $(A,t)_{EXT}$ is ε p-approximable in $Q'(1(a),\frac{1}{\varepsilon})$ time, choosing $\varepsilon = \frac{1}{k}$, $(A,t)_{EXT}$ is $\frac{1}{k}$ p-approximable in $Q'(1(a),k)$ time, and applying the same argument as in Theorem 4 we see that $(A,t)_{EXT,k}$ is recognizable in $Q'(1(a),k)$ time (that is: $Q = Q'$).

QED

REMARK. If $P \neq NP$ then p-simplicity is not a sufficient condition for fully p-approximability, as can be shown by the following NPOP "modified MAX SUBSET SUM": $(IS,t'_{ss})_{MAX}$, which is similar to max subset sum(*) with one exception: For an integer sequence $(a_1,a_2,\ldots,a_n,a_{n+1})$, t'_{ss} is defined by:

$$t'_{ss}((a_1,\ldots,a_{n+1}))=\{k|k \text{ divides } a_{n+1}, \text{ and } k = \sum_{j=1}^{m} a_{i_j} \text{ for some sequence } 1 \leq i_1<\ldots< i_m \leq n\}.$$

If k divides a_{n+1} and $k \neq a_{n+1}$, then $\frac{(a_{n+1})-k}{k} \geq 1$. That means that for $\varepsilon < 1$ any ε approximation for this problem will solve the NP complete Knpsack problem ([Ka 72]). Hence, if $P \neq NP$ then this problem is not p-approximable, and of course

(*) See Appendix.

not fully p-approximable; but it can be shown that this problem is p-simple (see next section).

DEFINITION 11. Let δ: Z → Z be a (recursive) function and let $(A,t)_{EXT}$ be a NPOP. Then:

$$(A,t)_{EXT,\delta(n)} = \{a \in A | op(a) \leq \delta(1(a))\}.$$

Let $(A,t)_{EXT}$ be a NPOP and let A' ⊂ A. Then the NPOP induced by $(A,t)_{EXT}$ on A' is the NPOP $(A',t)_{EXT}$. (It is assumed that A' is polynomial time recognizable).

Example. Let G' be the set of all planar graphs. Then the colorability problem induced on G' is the colorability problem for planar graphs.

The following lemma introduces a useful tool for recognizing p-simple NPOP's:

LEMMA 9. $(A,t)_{EXT}$ is p-simple implies that for any given polynomial $p_1(n)$, the NPOP induced by $(A,t)_{EXT}$ on $(A,t)_{EXT,p_1(n)}$ (namely the NPOP $((A,t)_{EXT,p_1(n)},t)_{EXT}$) is polynomially solvable. (Note that the lemma asserts two things: (a) that the problem induced on $(A,t)_{EXT,p_1(n)}$ is a NPOP (hence $(A,t)_{EXT,p_1(n)}$ is a set in P); (b) that this NPOP is polynomial solvable .

PROOF. $(A,t)_{EXT}$ is p-simple implies that there exists a (non-decreasing) polynomial Q(x,y), such that for all k ∈ Z, $(A,t)_{EXT,k}$ is recognizable in Q(1(a),k) time. By using binary search (see Theorem 3, pp. 6) one can show that the problem induced by $(A,t)_{EXT}$ on $(A,t)_{EXT,k}$ can be solved in Q'(1(a),k) time, where $Q'(1(a),k) \leq 0\ [Q(1(a),k) \log_2 k]$. This implies that $((A,t)_{EXT,p_1(n)},t)_{EXT}$ is solvable in Q'(1(a), $P_1(1(a)) = P_1(\ell(a))$ time, where P is a polynomial. QED

COROLLARY. Each of the following simple NPOP's cannot be fully p-approximable if $P \neq NP$.

(1) SET COVER
(2) MAX CLIQUE
(3) DOMINATING SET
(4) NODE COVER
(5) FEED BACK ARC SET
(6) FEED BACK NODE SET
(7) MAX SAT
(8) STEINER TREE
(9) MAX CUT.

PROOF. By Theorem 8, in previous page, it is sufficient to show that each of the above problems is not p-simple.

By the preceding Lemma 9, that can be done by showing that for some polynomial $p(n)$, each of the above problems satisfies the following: let $(A,t)_{EXT}$ be one of those problems, then the problem induced by $(A,t)_{EXT}$ on $(A,t)_{EXT,p(n)}$ is NPOP hard (that is: a polynomial solution of it implies $P = NP$).

Choosing $p(n) = n$ it is easily checked that each of the problems (1)-(7) satisfies $(A,t)_{EXT,n} = A$, and therefore the problems induced on $(A,t)_{EXT,n}$ are equal to the original problems which are easily shown to be NPOP complete. As for the problems (8)-(9) it is known that both of them remain NPOP complete even if we restrict the weight function W to be $W(i,j) = 1$ for all $(i,j) \in A$, (for (9) see [GJS 74]), and with this restriction, $op(G) \leq 1(G)$ for all $G \in \mathcal{G}$. Moreover, in these cases the set $(A,t)_{EXT,n}$ is a NP complete set. Thus, $P \neq NP$ implies that the problems induced by these NPOP's on $(A,t)_{EXT,n}$ are not necessarily NPOP's, and hence are NPOP hard but not necessarily NPOP complete.

While the above NPOPs are simple but not p-simple, we will prove that "MAX SUBSET SUM" is p-simple. For other examples, the reader is referred to [Sa 76].

Let $a = (a,\ldots,a_n) \in (Z)^{n+1}$ be an input to "MAX SUBSET SUM". Then the following algorithm will solve the problem:

"is $a \in (A,t)_{EXT,k}$?" in $O(\ell(a)\cdot k)$ time units. The algorithm contains a variable "T" which is the set of all "feasible solutions", and at the end of the algorithm $T = t_{ss}(a)$,

1. begin;
2. $T \leftarrow \{0\}$, $i \leftarrow 1$;
3. For every a in T do begin;
4. If $k < a+a_i \leq b$ then halt and reject (comment $op(a) > k$);
5. If $a+a_i \leq k$ then $T \leftarrow T \cup \{a+a_i\}$ end;
6. If $i = n$ then halt and accept;
7. $i \leftarrow i+1$ go to 2;
8. End.

The algorithm checks, for given a, whether $a \in (A,t)_{EXT,k}$ for this NPOP and its time complexity is $O(n\cdot k)=O(1(a)\cdot k)$. This follows from the fact that $|T| \leq k$ all through the execution of the algorithm.

4.3. *Sufficient Conditions for Fully p-Approximability*

We now introduce a property of NPOP's, that together with p-simplicity provides a sufficient (but not necessary) conditions for fully p-approximability.

DEFINITION 12. $(A,t)_{EXT}$ is condensable iff there exists a polynomial $E(n)$ and a polynomial time function $d:A\times Z \rightarrow A$, denoted by $d(a,c) = a_c$, that satisfies the following:

$$op(a_c) \leq \frac{op(a)}{c} \leq op(a_c)+E(1(a)) \text{ if } EXT = MAX \tag{4.3.1}$$

$$op(a_c) \geq \frac{op(a)}{c} \geq op(a_c)-E(1(a)) \text{ if } EXT = MIN. \tag{4.3.1'}$$

$E(n)$ will be referred to as the "*error function*".

Example. The TSP problem is condensable, where for a given weighted graph $W(G)$, $d(W(G),c)$ is the same graph G with a new wieght function Wc, defined by $Wc(i,j) = \frac{\lceil W(i,j) \rceil}{c}$ (*). $E(n)=n$ is a proper error function

THEOREM 10. Let $(A,t)_{EXT}$ satisfy the following:

(1) $(A,t)_{EXT}$ is p-simple;

(2) $(A,t)_{EXT}$ is condensable.

Then $(A,t)_{EXT}$ is fully p-approximable.

PROOF. We assume EXT = MAX, the proof is the same for EXT = MIN with minor changes. $(A,t)_{EXT}$ is p-simple $\rightarrow$ there exists a polynomial $Q'(x,y)$ (which is nondecreasing in both its variables) such that the NPOP induced by $(A,t)_{EXT}$ on $(A,t)_{EXT,k}$ is solvable in $Q'(1(a),k)$ time. In particular $(A,t)_{EXT,k}$ is recognizable in $Q'(1(a),k)$ time.
$(A,t)_{EXT}$ is condensable $\rightarrow$ there exists a polynomial $E(n)$ and a function : $A \times Z \rightarrow A$ as in Definition 12.

Let ε be given, for simplicity we assume $\varepsilon = \frac{1}{n}$ for some $n \geq 1$ so that $\frac{k}{\varepsilon}$ is an integer for every integer k. On input $a \in A$, we perform the following algorithms which contain 3 variables: "c" is an integer of the form 2^n. "b" is a word, $b = a_c$, and h is the ε approximation of op(a).
Begin

(1) $c \leftarrow 1$, $b \leftarrow a$, $\varepsilon' \leftarrow \varepsilon/2$;

(2) if $b \in (A,t)_{EXT,\ \frac{2}{\varepsilon'} \cdot E(\ell(a))}$ then go to (6);

(3) $c \leftarrow 2c$;

(4) $b \leftarrow a_c$

(*) For a real number r, $\lceil r \rceil$ stands for the least integer not smaller than r, and $\lfloor r \rfloor$ stands for the greatest integer not greater than r.

(5) go to (2);

(6) find $op(a_c)$: $h \leftarrow c \cdot op(a_c)$ end.

We have to prove two facts: that h is an ε-approximation to op(a), and that the algorithm requires $p(1(a),\frac{1}{\varepsilon})$ time for some polynomial p.

Case (a): $op(a) \le \frac{2}{\varepsilon'} E(1(a))$. In this case the algorithm finds $h = op(a)$ in $Q'(1(a),\frac{2}{\varepsilon'} E(1(a))) = p(1(a),\frac{1}{\varepsilon})$ time.

Case (b): $op(a) > \frac{2}{\varepsilon'} E(1(a))$. Let $c_{\varepsilon'}$ be first the value of the variable c when $b = (a_{c_{\varepsilon'}}) \in (A,t)_{EXT,\ \frac{2}{\varepsilon'}E(1(a))}$

Since $\frac{a_{c_{\varepsilon'}}}{2} \notin (A,t)_{EXT,\frac{2}{\varepsilon'}E(1(a))}$ we have that

$$\frac{2}{\varepsilon'}E(1(a)) < op(a_{\frac{c_{\varepsilon'}}{2}}).$$

On the other hand it follows from 4.3.1 that

$$op(a_{\frac{c_{\varepsilon'}}{2}}) \le \frac{2op(a)}{c_{\varepsilon'}} \le 2op(a_{c_{\varepsilon'}}) + 2E(1(a)).$$

Combining those inequalities we get that

$$\frac{1}{\varepsilon'}E(1(a)) < op(a_{c_{\varepsilon'}}) + E(1(a))$$

or

$$(\frac{1-\varepsilon'}{\varepsilon'})E(1(a)) < op(a_{c_{\varepsilon'}})$$

or

$$\frac{E(l(a))}{op(a_{c_{\varepsilon'}})} < \frac{\varepsilon'}{1-\varepsilon'} \leq 2\varepsilon' = \varepsilon.$$

Using again 4.3.1 for ε' and applying the above inequality we get

$$1 \leq \frac{op(a)}{c_{\varepsilon'} op(a_{c_{\varepsilon'}})} \leq 1 + \frac{E(l(a))}{op(a_{c_{\varepsilon'}})} < 1 + \varepsilon ;$$

or

$$0 \leq \frac{op(a) - c_{\varepsilon'} op(a_{c_{\varepsilon'}})}{c_{\varepsilon'} op(a_{c_{\varepsilon'}})} < \varepsilon ,$$

which shows that $h(=c_{\varepsilon'} op(a_c))$ is an ε-approximation to op(a). As for the timing: since there exists a polynomial $P_{2(n)}$ such that for all $a \in A$, $op(a) \leq 2^{P_2(l(a))}$, the loop (2)-(5) should be repeated no more than $P_2(l(a))$ time. The execution of lines (2) and (6) require no more than $Q'(\ell(a), \frac{2}{\varepsilon} \cdot E(\ell(a))) = p_1(l(a), \frac{1}{\varepsilon})$ time each.

The execution of line (4) requires $p_3(l(a))$ time for some polynomial p_3. Hence, the total time required for the execution of the algorithm is

$$0\{p_2(l(a))[p_1(l(a), \tfrac{1}{\varepsilon}) + p_3(l(a))]\} = p(l(a), \tfrac{1}{\varepsilon}$$

where p is the required polynomial. QED

REMARK. In practice the above algorithm can be greatly improved, since for most problems one can easily get a good estimation to op(a), instead of exhaustingly repeating the loop (2)-(5). (In "MAX SUBSET SUM", for instance, it can be shown that for an input $a = (a_1 \ldots a_n, a_{n+1})$, except for some trivial cases, $\frac{a_{n+1}}{2} < op(a) \leq a_{n+1}$).

Although the p-approximation technique introduced in Theorem 10 is not the most general one (see Remark, p.31) it holds for most of the *natural* NPOP's which have been shown to be fully p-approximable. Using Theorem 8 (for necessary conditions) and 10 (for sufficient conditions) one can show that for most of the *natural* NPOP's: Either the problem cannot be fully p-approximable (if $P \neq NP$) or it is fully p- approximable. There is at least one exception, as far as we know - which is the k-Chinese Postman problem ([FHK 76]): $(W(G) \times Z, t_{cp})_{MIN}$: where for a weighted graph W(G) and an integer $k, t_{cp}(W(G),k) = \{c \mid$ there exists k cycles $c_1, c_2 \ldots c_k$ that cover all edges of G, and $\max_{1 \leq j \leq k} W(c_j)\} = c$.

This problem is not known to be fully p-approximable, and so far we cannot prove that it *cannot* be fully p-approximable.

It can be shown that this problem is fully p-approximable <-> the NPOP induced by this problem of the class of weighted graphs with unit weights ($W(e) = 1 \ \forall e \in E$), is polynomially solvable.

We now give an analogue of Theorem 10 for p-approximation:

DEFINITION 13. $(A,t)_{EXT}$ is *strongly condensable* iff it is condensable with a constant error function (that is: $E(n) = E_0$ for some constant E_0).

THEOREM 11. Let $(A,t)_{EXT}$ satisfy the following:

(1) $(A,t)_{EXT}$ is simple;

(2) $(A,t)_{EXT}$ is strongly condensable.

Then $(A,t)_{EXT}$ is p-approximable. The proof is similar to that of Theorem 11, and is omitted.

Although the condition (1) of Theorem 11 is weaker than the corresponding condition of Theorem 10, which would be expected for getting a weaker result, still, the second condition of this theorem is much stronger than the corresponding condition of Theorem 10. This provides perhaps, some intuitive explanation to the fact that all *natural* NPOPs that are known to be p-approximable are in fact fully p-approximable .

REMARK. Further results concerning (fully) p-approximability, such as:

(a) A necessary *and* sufficient condition for (fully) p-approximability;

(b) Some fully p-approximable NPOP's, that require approximation technique which differ from that of Theorem 10,

will appear in a later paper.

BIBLIOGRAPHY

[AHU 74] A.V. AHO, J.E.HOPCROFT and J.D.ULLMAN: *The Design and Analysis of Computer Algorithms*, Addison Wesley (1974).

[Co 71] S.A.COOK: *The Complexity of Theorem Proving Procedures*, 3-rd STOC (1971), pp. 151-158.

[FHK 76] G.N.FREDERICKSON, M.S.HECHT and C.E.KIM: *Approximation Algorithms for some Routing Problems*, 17-th Annual Symposium on Foundations of Computing Sciences, pp. 216-227.

[GHS 74] M.R.GAREY, D.S.JOHNSON and L.STOCKMEYER: *Some Simplified NP-Complete Problems*, 6-th STOC (1974), pp. 47-63.

[HB 76] J.HARTMANIS and L.BERMAN: *On Isomorphism and Density of NP and Other Complete Sets*, 8-th STOC (1976), pp. 30-40.

[Jo 73] D.S.JOHNSON: *Approximation Algorithms for Combinatorial Problems*, 5-th STOC (1973), pp.38-49.

[Ka 72] R.M.KARP: *Reducibility among Combinatorial Problems*, R.E.Miller and J.W.Thatcher (eds.), Plenum Press, N.Y. (1972), pp. 85-104.

[Kn 74] D.E.KNUTH: *Postcript about NP Hard Problems*, SIGAT News, 23, (April 1974) pp. 15-16.

[PS 76] C.H.PAPADIMITRIOU and K.STEIGLITZ: *Some Complexity Results for the TSP*, 8-th STOC (1976), pp. 1-9.

[Sa 76] S.SAHNI: *General Techniques for Combinatorial Approximation*, TR. 76-6, University of Minnesota, Dept. of Computer Sciences, (1976).

APPENDIX

The following NPOP's were mentioned in the paper, but were not formally defined:

(1) TSP (Travelling Salesman Problem): $=(W(G), t_{TSP})_{MIN}$, where $W(G)$ is the set of all weighted graphs $W(G)$, (that is graphs combined with a weight function $W:A \to Z^+$), and for a given weighted graph $W[G(N,A)]$, $t_{TSP}(W[G(N,A)]) = \{k \mid$ there exists a Hamiltonian cycle in the graph whose weight is $k\} \cup \{\pm\infty\}$ (we add $\pm\infty$ to $t_{TSP}(W[G(N,A)])$ to make sure that it is not empty).

(2) MAX CUT: $= (W(G), t_{CUT})_{MAX}$, where $W(G)$ is as above and $t_{CUT}(W[G(N,A)] = \{k \mid A$ contains a cutset of weight $k\}$.

(3) MAX SUBSET SUM $= (IS, t_{ss})_{MAX}$ where $IS=\{(a_1, \ldots, a_n, a_{n+1})\}$ is the set of all finite integer sequences, and $t_{ss}((a_1, \ldots, a_n, a_{n+1})) = \{k \mid k \leq a_{n+1}$ and there are $1 \leq i_1 < \ldots < i_s \leq n$, $\sum_{j=1}^{s} a_{i_j} = k\}$.

(4) JSD (Job Sequencing with Deadlines) $= (IS^3, t_{JS})_{MAX}$ where:

$IS^3=\{T_1, D_1, P_1, \ldots, T_n, D_n, P_n) \mid \{T_i, D_i P_i\} \subset Z$ for $i=1, \ldots, n\}$,

and

$t_{JS}((T_1, D_1, P_1, \ldots, T_n, D_n, P_n))=\{k \mid$ there is a permutation σ of $\}$ $(1,2,\ldots,n)$

such that

$$\sum_{i=1}^{n} \delta_{\sigma(i)} P_{\sigma(i)} = k,$$

where

$\delta_{\sigma(i)} =[$if $T_{\sigma(1)} + T_{\sigma(2)} + \ldots + T_{\sigma(i)} \geq D_{\sigma(i)}$ then 0 else 1].

(5) SET COVER = $(f,t_{sc})_{MIN}$ where: f is the set of all finite families of finite sets, and for $\{S_1,\ldots,S_n\} \in f$
$t(\{S_1,\ldots,S_n\}) = \{i \mid \text{there exists } 1 \le j_1<j_2<\ldots<j_i \le n$ so that $\bigcup_{r=1}^{i} S_{j_r} = \bigcup_{r=1}^{n} S_r\}$.

(6) DOMINATING SET = $(G,t_{DS})_{MIN}$, where for $G \in G$,

$t_{DS}(G) = \{k \mid$ there are k nodes in G that are adjacent to all other nodes of G$\}$.

(7) CLIQUE COVER: $(G,t_{cc})_{MIN}$, where for $G(V,E) \in G$:

$t_{cc}(G) = \{k \mid$ there exists k cliques in G whose union is V$\}$.

(8) FEEDBACK ARC SET: $(D,t_{FBA})_{MIN}$, where D is the set of all directed graphs, and for $D(V,E) \in D$, (V=the set of vertices, E = the set of edges)

$t_{FBA}(D) = \{k \mid$ there exists k edges in A such that each directed) cycle in D contains at least one of them$\}$.

(9) FEEDBACK NODE SET: $(D,t_{FBN})_{MIN}$, where for $D(V,E) \in D$:

$t_{FBN}(D) = \{k \mid$ there exists k vertices in N such that each (directed) cycle contains at least one of them$\}$.

(10) STEINER TREE: $((W(G),S),t_{STR})_{MIN}$, where (W(G),S) is the set of all weigheted graphs together with a given subset of the nodes of the graph. For a given element (W(G),S) of this set,

$t_{STR}((W(G),S)) = \{k \mid$ there exists a subtree of G that contains S, whose weight is k$\}$.

A CHARACTERIZATION OF REDUCTIONS AMONG COMBINATORIAL PROBLEMS

G.AUSIELLO (*), A.D'ATRI (**), M.PROTASI (***)

(*) Centro di Studio dei Sistemi di Controllo e Calcolo Automatici del CNR, Via Eudossiana, 18 00184 Roma.

(**) Istituto di Automatica dell'Università di Roma, Via Eudossiana, 18 00184 Roma.

(***) Istituto di Matematica dell'Università dell'Aquila, Via Roma, L'Aquila

ABSTRACT

In this paper we introduce the concept of convex optimization problem associated to an NP-complete set and we study the combinatorial properties of families of such problems. The notion of structure is introduced and through this notion approximability properties and a partial ordering of families of optimization problems are shown.

1. INTRODUCTION

Many papers have been devoted to the study of the class of NP-complete problems from different points of view. Our interest is particularly concentrated on precise points and questions. After the basic results found by Cook [4] and Karp [8], some researchers have tried to find out how much these problems are similar, and instead some other authors have wondered whether it was possible to find properties which showed how different two NP-complete problems can be.

As regards the first area, Hartmanis and Berman [6], prove that all known NP-complete problems are equal up to a polynomial time computable isomorphism. Furthermore they show that these isomorphisms can be defined in such a way that they preserve the multiplicity of the solutions. In contrast with these results there are some papers (Garey and Johnson [5], Sahni [10], Sahni and Gonzales [11], and so on) which introduce a differentiation among NP-complete problems with respect to the approximation properties. There exist some problems (e.g. job sequencing) in which we can obtain arbitrarily "good" approximate polynomial solutions, while this is impossible for other problems (e.g. graph colouring). The interest of all these results has induced us to deepen this kind of study.

In our work (whose first results were presented in [1] and [2]) we have studied basically three problems:

i) by which tools we can distinguish among classes of NP-complete problems and of their associated optimization problems;

ii) how the combinatorial properties of classes of optimization problems relate with their approximability properties;

iii) what kind of classifications among optimization problems can be introduced.

After giving in §2 the formalization of the concepts of optimization problem and of its associated combinatorial problem we tackle the first issue in §3 by introducing the notion of "structure" of an input element to an optimization problem. The definition of this concept and the study of its properties are particularly promising in the case of "convex" problems. In fact, in this case, we can introduce a natural ordering over classes of inputs and define structure preserving reductions among different optimization problems. §4 is then devoted to study structure preserving reductions and their properties and to show (point ii)) under what conditions two problems have similar approximability properties. Finally §5 gives a first answer to point iii) by showing several examples of structure preserving reductions and the partial ordering among classes of optimization problems based on structure preserving reductions and structural isomorphisms.

2. OPTIMIZATION PROBLEMS AND ASSOCIATED COMBINATORIAL PROBLEMS

Let us first briefly review the basic terminology and notation.

Let Σ^* be the set of all words over a finite alphabet Σ. A language $L \subseteq \Sigma^*$ is said to be recognizable in time $t(n)$ by a Turing Machine (TM) M if for all $n \geq 0$, for every input x of length n, M takes less than $t(n)$ steps either to accept or to reject x. If the TM is non deterministic we will consider the number of steps of the shortest accepting computation (if x is accepted) or the number of steps of the longest rejecting computation (if x is rejected).

DEFINITION 1. $NP = \{L \mid L$ is recognizable by a non-deterministic TM in time bounded by some polynomial $p\}$.

DEFINITION 2. A set $A \subseteq \Sigma^*$ is said to be p-*reducible to a set* $B \subseteq \Gamma^*$ (denoted $A \leq B$) if there is a mapping $f: \Sigma^* \rightarrow \Gamma^*$ which is computable in polynomial time on a deterministic TM and such that for every $x \in \Sigma^*$, $f(x) \in B$ if and only if $x \in A$.

DEFINITION 3. A set B is said to be *complete* for some

class of sets C (denoted C-complete) if $B \in C$ and, for every $A \in C$, $A \leq B$.

In the following example and throughout the paper we assume that all sets are encoded into Σ^* under some natural encoding.

Well known examples of NP-complete sets (problems) are:
SATISFIABILITY = {w|w is a formula of propositional calculus in CNF and there exists a truth assignment that satisfies it}.

CLIQUE = $\{\langle G,K\rangle \mid$ G is a graph, K is an integer and G has a complete subgraph of K nodes}.

CHROMATIC-NUMBER = $\{\langle G,K\rangle \mid$ G is a graph, K is an integer and G can be coloured with K colours with no two adjacent nodes equally coloured}.

DIOPH = $\{\langle a,b,c\rangle \mid a,b,c \geq 0$ are integers and the quadratic diophantine equation $ax^2 + by - c = 0$ can be solved with x,y positive integers}.

SUBSET-SUM = $\{\langle a_1,\dots,a_n,b\rangle \mid$ there is a subsequence $i_1,\dots,i_m$ such that $\sum_{j=1}^{m} a_{i_j} = b\}$.

JOB-SEQUENCING-WITH-DEADLINES = $\{\langle t_1,\dots,t_n, d_1,\dots,d_n, p_1,\dots,p_n,k\rangle \}$ there exists a permutation π such that $\sum_{j=1}^{n} (\textit{if} \sum_{i=1}^{j} t_{\pi(i)} \leq d_j \textit{ then } p_j \textit{ else } 0) \leq k$.

As it can easily be seen some of these sets are naturally related to optimization problems. In the following we will restrict ourselves to considering optimization problems of this type as it will be formally expressed in definition 6 below.

First of all let us give some definitions which are

derived from those given in [7]. In these definitions we try to capture the basic objects which constitute an optimization problem: essentially *input* objects, *output* objects, a subset of which is the *search space* for *approximate solutions*, a *measure* over the solutions, an ordering on the values of the measure which characterizes an optimization problem as a *maximization* or a *minimization* problem.

DEFINITION 4. An *NP optimization problem* (over an alphabet Σ) is the 5-tuple $A = \langle$ INPUT, OUTPUT, S, Q, m $\rangle$

INPUT and OUTPUT are infinite polynomially decidable subsets of Σ^*

S: INPUT $\rightarrow P(\Sigma^*)$ is a nondeterministic polynomial mapping that provides the search space for an input element x such that the set of all approximate solutions is given by the set of strings which are in S(x) and which belong to the output set. With the notation SOL we mean the set of approximate solutions, that is the set SOL(x) = S(x) $\cap$ OUTPUT

Q: is a countable totally ordered set

m: SOL(INPUT) $\rightarrow$ Q is the measure and is also polynomially computable.

For example if we consider the problem MIN-CHROMATIC-NUMBER we have:

INPUT: (encodings of) all undirected finite graphs;

OUTPUT: (encodings of) pairs $\langle G,P \rangle$ where G is a finite graph and P is a partition of the nodes of G such that (y,z) arc of G implies y and z in different classes of P

S(x) : (encodings of all) pairs $\langle x,T \rangle$ where T is a partition of the nodes of x

$Q^{(1)}$: integers in decrasing order

m : number of classes of P.

All the other optimization problems which we will study in this work, are defined in Appendix.

DEFINITION 5. i) The *optimal value* $m^*(x)$ of an input x of A is

$$m^*(x) = \text{best } \{m(y) \mid y \in SOL(x)\} \text{ under the ordering of } Q$$

ii) The *trivial value* $\tilde{m}(x)$ of an input x of A is

$$\tilde{m}(x) = \text{worst } \{m(y) \mid y \in SOL(x)\} \text{ under the ordering of } Q$$

DEFINITION 6. The *combinatorial problem* A^c *associated* to an optimization problem A is the set

$$A^c = \{\langle x,k \rangle \mid m^*(x) \geq k \text{ under the ordering of } Q\}$$

An interesting characterization of combinatorial problems associated to optimization problems is expressed in the following normal form result. (A different normal form result is given in [9]).

FACT. Let A^c be a combinatorial problem associated to an optimization problem A; then

$$A^c = \left\{\langle x,k \rangle \mid (\exists z \in S(x)) \left[z \in OUTPUT \text{ and } m(z) \geq k \text{ under the ordering of } Q \right] \right\}$$

(1) In all problems that will be considered in the text Q is the set of nonnegative integers under increasing or decreasing order according to the fact that the problem is a MAXmization or MINimization problem.

If A^c is in NP the Turing machine that recognizes it can be defined in the following way:

$$M = M_{IN} \cdot M_s \cdot M_{OUT} \cdot M_m$$

where, on input $w \in \Sigma^*$, M_{IN} rejects all the strings which are not of the form $\langle x,k \rangle$, M_s is a TM that in polynomial time non deterministically provides $z \in S(x)$ and M_{OUT}, M_m are deterministic TM which in polynomial time check whether $z \in OUTPUT$ and $m(z) \geq k$.

In this case if, given a problem A^c there is a polynomial q such that for every $x \in INPUT \| S(x) \| \leq q(|x|)$ A^c is in the class P.

3. ON THE STRUCTURE OF NPCO PROBLEMS

Our main interest is devoted to the study and classification of optimization problems whose associated combinatorial problem is NP-complete. We denote these problems *NPCO problems*. It should be noticed that not all NP-complete problems can be seen as being naturally associated to NP optimization problems. Besides, for some NP-complete problems, for some input elements every approximate solution is as hard as the optimal solution. This for example happens with the problem MIN-EXACT-COVER.
For this reason we will only consider those NPCO problems which admit a polynomial approximation, that is for which there exists a polynomially computable function f that, given any $x \in INPUT$, provides at least one element of SOL(x).

A first observation of the family of NPCO problems allows us to distinguish them in two classes according to the following general definition:

DEFINITION 7. An optimization problem is said to be

convex if m(OUTPUT) is the set of non-negative integers and for every $x \in$ INPUT, for every integer n between $\tilde{m}(x)$ and $m^*(x)$ there is at least one approximate solution of x whose measure is equal to n.

The relevant fact with convex optimization problems is that in practice, for such problems, for any input x, the existence of an approximate solution of measure q implies the existence of a certain number of approximate solutions for all measures between q and $\tilde{m}(x)$, while this is not the case with non convex problems. For example MIN-CHROMATIC-NUMBER is a convex problem while MAX-SUBSET-SUM is not.

Now let A be an NPCO problem; in the following definition we introduce the concept of structure of an element of INPUT, which is a fundamental concept in the development of our work.

DEFINITION 8. Let $x \in$ INPUT; we define *structure* of x the list $\ell_x = \langle a_0, \ldots, a_c \rangle$ where $a_i = \| \{ y \mid y \in SOL(x)$ and $i = |m(y) - \tilde{m}(x)| \} \|$ and $c = |m^*(x) - \tilde{m}(x)|$.

The concept of structure, which provides a representation of the number of approximate solutions at different levels of the measure, contains enough information to capture the combinatorial properties of the input to an optimization problem, and, as we will see, allows us to introduce an ordering over all input elements that expresses the concept of "subproblem".

For example, let A be the problem MIN-SET-COVERING; let x be the family of sets

$x = \{S_1, S_2, S_3, S_4, S_5\}$ where

$S_1 = \{1,2,3,4\}$ $S_2 = \{1,2,5\}$ $S_3 = \{3,4,6\}$ $S_4 = \{5,6,7\}$

$S_5 = \{1,2,3,4,5,6,7\}$

$\tilde{m}(x) = 5, \quad m^*(x) = 1, \quad \ell_x = \langle 1,5,9,5,1 \rangle .$

If we now consider the following example in which we make a change on the input element, such as eliminating one of the elements in the family, we have, essentially, a "subproblem" of the given problem: say

$y = \{S_1, S_2, S_3, S_4\},$

$\tilde{m}(y) = 4, \quad m^*(y) = 2, \quad \ell_y = \langle 1,3,1 \rangle .$

Clearly ℓ_y can be considered a "sublist" or ℓ_x. This concept is formalized in the following definitions.

DEFINITION 9. Let x and y be two input elements to a problem A and let $\ell_x = \langle a_0, \ldots, a_c \rangle$, $\ell_y = \langle b_0, \ldots, b_d \rangle$. We say that

i) $x \equiv y$ if ℓ_x coincides with ℓ_y ,

and, if [x] denotes the equivalence class of x,

ii) $[x] \leq [y]$ if for all $0 \leq i \leq \min\{c,d\}$ $a_i \leq b_i$.

On the base of definition 9, given a NPCO problem A, we can consider the set INPUT/$_\equiv$ (the set of equivalence classes over INPUT) as a partially ordered set.

If we refer to specific examples we can easily notice that the definition of equivalence given above is weaker than the intuitive concept of isomorphism. In [1] we observed this fact in the case of the problem MAX-CLIQUE. Here we can observe that in the case of MIN-SET-COVERING the following elements of INPUT

$$z = \Big\{\{2,3,4,5\},\{2,3,6\},\{4,5,7\},\{6,7,1\}\Big\}$$

$$u = \Big\{\{1,2,3,4,8\},\{1,2,5,8\},\{3,4,6,8\},\{5,6,7,8\}\Big\}$$

are such that [y] = [z] = [u] but while z is clearly "isomorphic" to y, u is equivalent to y while not being isomorphic.

This fact shows that the concept of structure is weak enough to capture not only the obvious isomorphism among input elements but also the similarity of their combinatorial properties.

The concept of structure and the consequent ordering over $\mathrm{INPUT}/_{\equiv}$, hence, are useful tools for the classification of NPCO problems according to their combinatorial properties. The first step in this direction is the definition of reductions among NPCO problems which preserve the structural properties.

DEFINITION 10. Given two NPCO problems A and B we say that *A is polynomially reducible to B* if there are two polynomial time computable functions f_1: $\mathrm{INPUT}_A \rightarrow \mathrm{INPUT}_B$, f_2: $\mathrm{INPUT}_A \times Q_A \rightarrow Q_B$ such that for every x and k

$$k \in m_A(\mathrm{SOL}_A(x)) \text{ iff } f_2(x,k) \in m_B(\mathrm{SOL}_B(f_1(x))).$$

Clearly if $\langle f_1,f_2 \rangle$ is a reduction between two optimization problems A and B, we have $\langle x,k \rangle \in A^c$ iff $\langle f_1(x),f_2(x,k)\rangle \in B^c$ and hence there exists a reduction between the corresponding combinatorial problems.

DEFINITION 11. A reduction $\langle f_1,f_2 \rangle$ from A to B is said to be

i) *order preserving* if, given $x,y \in \mathrm{INPUT}_A$, $[x] \leq [y]$ implies $[f_1(x)] \leq [f_1(y)]$

ii) *structure preserving* if for every $x \in \mathrm{INPUT}_A$ $\ell_x = \ell_{f_1(x)}$.

For example let us consider the problems MAX-CLIQUE and MAX-SET-PACKING. The reduction given in [8] is structure preserving (as it will be shown as a consequence of general results below).

Let x be the graph $\langle N, A \rangle$

$f_1(x) = \{S_1, \dots, S_n\}$ where $n = \|N\|$

$f_2(x,K) = K$ $\quad S_i = \Big\{ \{i,j\} \mid \{i,j\} \notin A \Big\}$

For example:

$$A_A = \Big\{ \{A,A\}, \{A,C\} \Big\}$$

$$S_B = \Big\{ \{B,B\} \Big\}$$

$$S_C = \Big\{ \{C,C\}, \{A,C\}, \{E,C\} \Big\}$$

$$S_D = \Big\{ \{D,D\} \Big\}$$

$$S_E = \Big\{ \{E,E\}, \{E,C\} \Big\}$$

$$\ell_x = \ell_{f_1(x)} = \langle 1,5,8,5,1 \rangle$$

Note that the only way the set S_U and the set S_V cannot be disjoint is if they both contain the element $\{U,V\}$. Hence, for every subgraph G' in x, for every arc $\{U,V\}$ in G' the corresponding sets S_U and S_V are disjoint and viceversa. This proves that for every K-clique in x there are K disjoint sets among $S_1, \dots, S_n$. Besides, if two K-cliques differ for at least one node, also the two corresponding families of K disjoint sets are different and viceversa. This proves that the number of solutions of measure K in the problem CLIQUE is the same as the number of solutions of measure K in the problem SET-PACKING. Since $f_2(x,K) = K$ we have that if in the list of x there are b solutions of measure u and c solutions of measure $v > u$ in the list of $f_1(x)$, b will still precede c. Hence we can conclude that the problem MAX-CLIQUE is polynomially, structure preserving reducible to MAX-SET-PACKING.

4. STRUCTURE PRESERVING REDUCTIONS

In the example given at the end of the preceding paragraph, the proof makes use of the following facts in order to prove the existence of a structure preserving reduction:

i) the reduction preserves the number of solutions at corresponding levels of the measures,

ii) the measures are related via a very simple monotonous function (the identity function).

A generalization of this observation is possible and brings the following results.

First of all we need a definition which is a translation into our terminology of the concept of parsimonious reduction (originally due to Simon [12] and Hartmanis and Berman [6]) but which also is a strengthening of the same concept.

DEFINITION 12. A reduction $f = \langle f_1, f_2 \rangle$ from A to B is said to be parsimonious if

$$(\forall x \in \mathrm{INPUT}_A)(\forall k \in Q_A)$$

$$\|\{y \in \mathrm{SOL}_A(x) \mid m_A(y)=k\}\| = \|\{y \in \mathrm{SOL}_B(f_1(x)) \mid m_B(y)=f_2(x,k)\}\|$$

Trivially it can be seen that our definition implies the one given in [12] but the viceversa is not true because of our definition of reduction between two optimization problems where the dependence on the argument is separated into the dependence on the input and the dependence on the measure.

We are interested in conditions for having structure preserving reductions. Even if the parsimoniousness of a reduction is an important condition for the reduction to be structure preserving it is not sufficient and we need the following theorem:

THEOREM 1. Let A and B be two convex NPCO problems; let $f = \langle f_1, f_2 \rangle$ be a reduction from A to B such that

i) f is parsimonious,

ii)
$$f_2(x,k) = \begin{cases} a(x) + k \text{ if } A \text{ and } B \text{ are both maximization (minimization) problems} \\ a(x) - k \text{ otherwise} \end{cases}$$

iii) f_2 is such that $f_2(x, \tilde{m}_A(x)) = \tilde{m}_B(f_1(x))$

then f is structure preserving.

PROOF. The fact that the problems A and B are convex and conditions i), ii) imply that for every a_i in the structure of $x \in INPUT_A$ there exists b_j in the structure of $f_1(x)$ such that $a_i = b_j$ and if moreover $f_2(x,k) = h$, where k,h are respectively the values of the measure corresponding to a_i and b_j, then we have that $a_{i+l} = b_{j+e}$ for every $l \geq 0$. Finally, because of condition iii) the above property is satisfied by a_0 and b_0 and the two lists coincide

QED.

Clearly the example given in the preceding paragraph falls in the conditions of theorem 1.

It can be noted, now, that the conditions given in theorem 1 are sufficient but not necessary to guarantee that the structures are preserved. In fact it can be seen that we

might have the following situation.
Let us consider the former example again and let x be an input element to the problem MAX-CLIQUE where x is a complete graph. in this case the structure of x is symmetric (if n is the order of the graph then

$\ell_x = \langle \binom{n}{0}, \binom{n}{1}, \ldots \binom{n}{n} \rangle$). Let $\langle f_1, f_2 \rangle$ be the reduction

from MAX-CLIQUE to MAX-SET-PACKING which was defined above. We can define a new reduction $\langle f_1', f_2' \rangle$ where $f_1' = f_1$ and

$$f_2'(x) = \begin{cases} n - f_2(x) & \text{if x is a complete graph of order n} \\ f_2 & \text{otherwise} \end{cases}$$

Clearly $\langle f_1', f_2' \rangle$ is still a polynomial structure preserving reduction from MAX-CLIQUE to MAX-SET-PACKING but condition ii) is not satisfied.

For this reason we may consider in the following result in what cases the conditions of theorem 1 are also necessary.

First we need the following definition:

DEFINITION 13. A reduction from A to B is said to be *strictly monotonous* if for every x, and for every $\tilde{m}(x) \leq k_1, k_2 \leq m^*(x)$ $k_1 < k_2$ in Q_A implies $f_2(x,k_1) < f_2(x,k_2)$ in Q_B.

Then we may prove:

THEOREM 2. Let A and B be two convex NPCO problems. Let $f = \langle f_1, f_2 \rangle$ be a strictly monotonous, structure preserving reduction from A to B. Then $\langle f_1, f_2 \rangle$ must satisfy the conditions of theorem 1.

PROOF. First of all let us notice that since f is structure preserving, it preserves also the length of the lists.

Now, from the fact that f is length preserving and strictly monotonous we derive condition iii), that is $f_2(x,\tilde{m}_A(x)) = \tilde{m}_B(f_1(x))$. Besides, the fact that the problems are convex plus the fact that f is strictly monotonous and length prescriving and the already achieved condition iii) imply that f_2 must be as defined in condition ii).
Finally, from condition ii) and the fact that f is structure preserving we infer that f is parsimonious.

QED

As we will see, theorem 1 is powerful enough to prove that a large number of reductions given in the literature are indeed structure preserving. The concept of structure preserving reduction is sufficiently weak to capture the combinatorial similarity among classes of combinatorial problems. Such power, actually, cannot be achieved by using other types of reductions such as the "measure preserving" reductions [9], though the last one has been shown very useful to discuss the possibility of preserving "good" approximations.

Note that, on the other side, we may find simple but interesting relations among the existence of structure preserving reductions and the possibility of preserving the degree of approximability.

DEFINITION 14. Let A be an NPCO problem. We say that any algorithm A that maps INPUT_A into OUTPUT_A is an *approximate algorithm* for A if $(\forall x \in \text{INPUT}_A)[\text{A}(x) \in \text{SOL}(x)]$.

DEFINITION 15. Given an NPCO problem A and an approximate algorithm A we define *proximity degree on input* x the value

$$r_A(x) = \frac{m^*(x)-m(\text{A}(x))}{m^*(x)-\tilde{m}(x)}$$

DEFINITION 16. Let A be an NPCO problem. We say that A is *polynomially approximable* if, given any $\varepsilon > 0$ there exists a polynomial approximate algorithm A_ε such that the proximity degree $r_{A_\varepsilon}(x)$ is bounded by ε.

THEOREM 3. Let A and B be two convex NPCO problems. If there exist two reductions $f = \langle f_1, f_2 \rangle$ from A to B and $g = \langle g_1, g_2 \rangle$ from B to A such that

i) both are structure preserving

ii) both are strictly monotonous

iii) $f_2(x,k) = a(x)+k$, $g_2(y,h) = b(y)+h$ and $a(x) \geq -b(f_1(x))$ if the problems are both maximization or minimization problems or,

iii)' $f_2(x,k) = a(x)-k$, $g_2(y,h) = b(y)-h$ and $a(x) \leq b(f_1(x))$ otherwise,

then if B is polynomially approximable, so is A, and vice-versa.

PROOF. First of all notice that the first part of conditions iii) and iii)' is already implied by conditions i) and ii) according to theorem 2. Let us first consider the case that condition iii) holds.

Now, given $\varepsilon > 0$, suppose A_ε is the ε-approximate algorithm for problem B, then it is true that, given $x \in \mathrm{INPUT}_A$

$$\alpha(x) = \frac{m_B^*(f_1(x)) - m_B(A_\varepsilon(f_1(x)))}{m_B^*(f_1(x)) - \tilde{m}_B(f_1(x))} \leq \varepsilon$$

We can show that, given any $x \in \mathrm{INPUT}_A$, via the mapping f_1, the approximate algorithm A_ε for B, and the mapping g_2, we may find the measure of an approximate solution of x whose proximity degree, with respect to the optimal solution of x, is bounded by ε. In fact we have that the said proximity degrees is:

$$\beta(x) = \frac{m_A^*(x) - g_2(f_1(x), m_B(A_\varepsilon(f_1(x))))}{m_A^*(x) - \tilde{m}_A(x)} =$$

$$= \frac{m_A^*(x) - (b(f_1(x)) + m_B(A_\varepsilon(f_1(x))))}{m_A^*(x) - \tilde{m}_A(x)}$$

On the other side, notice that

$$\alpha(x) = \frac{a(x) + m_A^*(x) - m_B(A_\varepsilon(f_1(x)))}{a(x) + m_A^*(x) - (a(x) + \tilde{m}_A(x))}$$

By comparing the two expressions we can see that since $a(x) \geq -b(f_1(x))$ then $\alpha(x) \leq \beta(x) < \varepsilon$.
The analogous proof can be carried on for the case that condition iii)' holds.

QED

In the next paragraph we will see some examples of classes of convex NPCO problems that satisfy the hypotheses of theorem 3.

5. CLASSES OF STRUCTURALLY ISOMORPHIC NPCO PROBLEMS

In this paragraph essentially applying the results given in theorem 1, we give an ordering and then a classification of convex NPCO problems.

Let us consider the class of all convex NPCO problems; an ordering over the class can be introduced according to the following:

DEFINITION 17. Given two convex NPCO problems A and B we say that

i) $A \leq_{sp} B$ if there exists a polynomial structure preserving reduction from A to B;

ii) $A \equiv_{sp} B$ if $A \leq_{sp} B$ and $B \leq_{sp} A$; in this case we say that A and B are *structurally isomorphic*;

iii) $A \not\leq_{sp} B$ if no structure preserving reduction is possible from A to B;

iv) $A \mid<_{sp} B$ if $A \leq_{sp} B$ and $B \not\leq_{sp} A$

v) $A \between_{sp} B$ if $A \not\leq_{sp} B$ and $B \not\leq_{sp} A$

The above definitions characterize the fact that given two optimization problems it may be i) that one of them is at most as rich as the other one with respect to the combinatorial structure and we have a polynomial reduction among them, ii) that they have exactly the same combinatorial structure and polynomial mappings in both direction can be exhibited, iii) that some structures that are present in the first one cannot be found in the second one, iv) that one of them is strictly richer than the other one, v) that their combinatorial structures are incomparable.

According to definition 17, we give the following results:

THEOREM 4. MAX-CLIQUE $\equiv_{sp}$ MIN-NODE-COVER $\equiv_{sp}$ MAX-SET-PACKING

PROOF. The results MAX-CLIQUE $\leq_{sp}$ MIN-NODE-COVER and MAX-CLIQUE $\leq_{sp}$ MAX-SET-PACKING can be proved, considering the reductions given in [8], which can be easily seen to verify the hypotheses of theorem 1. Applying the inverse reduction we have that MIN-NODE-COVER $\leq_{sp}$ MAX-CLIQUE.
Finally we can give a reduction from MAX-SET-PACKING to MAX-CLIQUE in which $f_1(x)$ is a graph where the nodes correspond to the sets in x and the arcs correspond to pairs of disjoint sets, and $f_2(x,k) = k$; clearly this reduction

satisfies theorem 1.

QED

THEOREM 5. MIN-SET-COVERING $\equiv_{sp}$ MIN-HITTING-SET

PROOF. $\leq_{sp}$) Let $x = \{S_1,\dots,S_n\}$ be an input to MIN-SET-COVERING; $f_1(x) = \{U_1,\dots,U_n\}$ where $s_j \in U_i$ iff $u_i \in S_j$; $f_2(x,k) = k$
By theorem 1, we can prove the result

$\geq_{sp}$) It is sufficient to consider the inverse reduction

QED

THEOREM 6. MIN-CHROMATIC-NUMBER $\equiv_{sp}$ MIN-EXACT-CLIQUE-COVER

PROOF. $\leq_{sp}$) If we consider the reduction obtained from the reduction given in [8], from CHROMATIC-NUMBER to CLIQUE-COVER, we can apply theorem 1.

$\geq_{sp}$) The inverse reduction can be easily proved to be a structure preserving reduction

QED

Notice that, because MIN-CHROMATIC-NUMBER is not polynomially approximable [5], applying theorem 3 and 6, we could easily prove MIN-EXACT-CLIQUE-COVER is not polynomially approximable.

THEOREM 7. i) MIN-NODE-COVER $|<$ MIN-SET-COVERING

ii) MIN-NODE-COVER $|<$ MIN-FEEDBACK-NODE-SET

PROOF. i) $\leq_{sp}$) the reduction given in [8] from MIN-NODE-COVER to MIN-SET-COVERING satisfies theorem 1

$\not\geq_{sp}$) let us consider the family of sets $x = \{S_1,\dots,S_n\}$ where $S_i = A-a_i$ and

$A = \{a_1, \ldots, a_n\}$ with $n \geq 3$.

In this case $\ell_x = \langle 1, \binom{n}{n-1}, \binom{n}{n-2}, \ldots, \binom{n}{2} \rangle$ and such a list does not exist in the problem MAX-CLIQUE because if $\ell_x = \langle 1, \binom{n}{n-1}, \binom{n}{n-2}, \ldots \rangle$ x must be an n-clique and hence its list must be $\langle 1, \binom{n}{n-1}, \binom{n}{n-2}, \ldots, \binom{n}{2}, n, 1 \rangle$.

By theorem 4 the same list ℓ_x does not exist in the problem MIN-NODE-COVER.

ii) $\leq_{sp}$) the reduction given in [8] satisfies theorem 1

$\geq_{sp}$) let us consider the family of graphs formed by just one cycle on n nodes. For such a graph the list in MIN-FEEDBACK-NODE-SET would be

$$\langle 1, \binom{n}{n-1}, \binom{n}{n-2}, \ldots, \binom{n}{2}, n \rangle$$

and such a list does not exist in MIN-NODE-COVER for the same reason as in part i).

QED

THEOREM 8. MIN-FEEDBACK-ARC-SET $|\leq_{sp}$ MIN-FEEDBACK-NODE-SET

PROOF. $\leq_{sp}$) Let $G = \langle V, E \rangle$ be a digraph. Let us consider the digraph $G' = \langle V', E' \rangle$ where $V' = E$ and $E' = \{\langle e, e' \rangle \mid (\exists x, y, v \in V)[e = \langle x, v \rangle, e' = \langle v, y \rangle]\}$. In the case that G has no selfloops we have a 1-1 correspondence between all cycles in G and all cycles in G' so that to any distinct feedback

arc cover in G there is a corresponding distinct feedback node cover in G'.
In the case that G has selfloops the correspondence between cycles fails but the correspondence between coverings is still preserved.

$\geq_{sp}$) Let us consider the family of complete digraphs of n nodes without selfloops. These digraphs have the list $\langle 1,n \rangle$ in MIN-FEEDBACK-NODE-SET. For $n \geq 3$ this list cannot be found in MIN-FEEDBACK-ARC-SET because this would imply that any pair of arcs is itself a loop and this is obviously impossible

QED

Note that in some cases we might prove that the structures that can be found in one NPCO problem are also present in another NPCO problem, that is for any $x \in INPUT_A$ there is $y \in INPUT_B$ such that $\ell_x = \ell_y$, even if we are unable of exhibiting a structure preserving reduction from A to B that given x computes such a y in polynomial time. In this case we will write $A \subseteq B$. For example we have

THEOREM 9. MIN-FEEDBACK-NODE-SET $\subseteq$ MIN-SET-COVERING.

PROOF. Let $G = \langle V,E \rangle$ be a digraph. We may define the following family of sets $[A_1,\ldots,A_n]$ where n is the number of vertices in V, $\cup A_i = \{u_1,\ldots,u_m$ where m is the number of cycles in G$\}$ and, for every i and $j, u_j \in A_i$ if and only if the i-th node is in the j-th cycle. Note that since the number of nodes, this construction cannot be used as a polynomial reduction from MIN-FEEDBACK-NODE-SET to MIN-SET-COVERING, but it is sufficient to show that every structure present in the first problem can be found in the second problem.

QED

As a last remark we may notice that in the definition of MIN-SET-COVERING we may require the input to be a "set" instead of "family") of sets because the following is true:

FACT. MIN-SET-COVERING $\equiv_{sp}$ MIN-UNI-SET-COVERING

$\geq_{sp}$) Obvious.

$\leq_{sp}$) Let $x = [A_1, \ldots, A_n]$ be a multiset of sets. If for no i and j $A_i = A_j$ clearly the problem becomes an uniset covering problem. Otherwise, let us suppose to have $m \geq 1$ families of equal sets. Let $[A_1, \ldots, A_k]$ be such a family, that is $A_1 = \ldots = A_k$ and $A_{k+1}, \ldots, A_n \neq A_1$. We will show that there is another element y of MIN-SET-COVERING which has exactly m-1 families of equal sets and $\ell_x = \ell_y$.
Such an element is $y = [B_1, \ldots, B_k, C_{k+1}, \ldots, C_n, D_{n+1}]$ where $B_i = A_i \cup \{z_i\}$ $1 \leq i \leq k$

$$C_i = A_i \cup \{z_1, \ldots, z_k\} \quad k+1 \leq i \leq n$$

$$D_{n+1} = \{z_1, \ldots, z_k, z_{k+1}\}$$

where $z_1, \ldots, z_{k+1}$ are new elements not in $\cup A_i$.
By induction the fact is proved.

QED

APPENDIX

MAX-CLIQUE

INPUT : finite graphs

OUTPUT : complete finite graphs

S(x) : subgraphs of a given graph x

m : number of nodes of a complete graph

MIN-NODE-COVER

INPUT : finite graphs

OUTPUT : pairs $\langle G,N \rangle$, where G is a finite graph and N is a set such that for any arc $\langle z,y \rangle$ of G, $z \in N$ or $y \in N$

S(x) : pairs $\langle x,M \rangle$ where M is a subset of nodes of x

m : number of nodes of M

MAX-SET-PACKING

INPUT : families of finite sets

OUTPUT : families of mutually disjoint finite sets

S(x) : subfamilies S' of x

m : number of sets in the subfamily S'

MIN-SET-COVERING

INPUT : pairs $\langle F,A \rangle$ where F is a family of finite sets and A is a finite set such that $\bigcup_{S_i \in F} S_i \subseteq A$

OUTPUT : pairs $\langle H,B \rangle$ where H is a family of finite sets and B is a finite set such that $\bigcup_{H_i \in H} H_i = B$

S(x) : pairs $\langle F',A' \rangle$ where F' is a subfamily of F and $A' = \bigcup_{S_i \in F} S_i$

m : number of sets of F'

MIN-EXACT-COVER : as MIN-SET-COVERING where F' is a family of disjoint sets.

MIN-HITTING-SET:

INPUT : families of finite sets

OUTPUT : pairs $\langle F,T \rangle$ where F is a family of sets and S is a set such that $(\forall S_i \subseteq F)[\|T \cap S_i\| \geq 1]$

S(x) : pairs $\langle x,S \rangle$ where $S \subseteq \bigcup_{S_i \in x} S_i$

m : number of sets of S

MIN-EXACT-CLIQUE-COVER:

INPUT : finite graphs

OUTPUT : pairs $\langle F,H \rangle$ where F is a family of complete disjoint finite graphs $\langle N_i,A_i \rangle$, H is a finite set such that $\cup N_i = H$

S(x) : pairs $\langle F',N \rangle$ where F' is a set of subgraphs of x and N is the set of nodes of x

m : number of subgraphs in F

MIN-FEEDBACK-NODE-SET (MIN-FEEDBACK-ARC-SET)

INPUT : finite digraphs $\langle V,E \rangle$

OUTPUT : pairs $\langle G,A \rangle$ where G is a finite digraph and A is a set of vertices (edges) such that all cycles in G have a vertex (edge) in A

S(x) : pairs $\langle x,A \rangle$ where A is a subset of the set of vertices (edges) of x

m : number of vertices (edges) in A

MAX-SUBSET-SUM

INPUT : pairs $\langle A,b \rangle$ where A is a finite set of integers and b is an integer

OUTPUT : pairs $\langle C,d \rangle$ where $\sum_{c_i \in C} c_i \leq d$ (c_i, d are integers)

S(x) : pairs $\langle A',b \rangle$, where A' is a subset of A

m : $\sum_{a_i \in A'} a_i$

BIBLIOGRAPHY

[1] G.AUSIELLO, A.D'ATRI, M.PROTASI: *On the structure of combinatorial problems and structure preserving reductions.* Proc. 4th Int. Coll. on Automata, Languages and Programming Turku, 1977.

[2] G.AUSIELLO, A.D'ATRI, M.GAUDIANO, M.PROTASI: *Classes of structurally isomorphic NP-optimization problems.* Proc. 6th Symp. on Mathematical Foundations of Computer Science, Tatranska Lomnica, 1977.

[3] L.BERMAN, J.HARTMANIS: *On polynomial time isomorphisms of complete sets.* Proc. 3rd GI Conference, Darmstadt, 1977.

[4] S.COOK: *The complexity of theorems proving procedures.* Proc. 3rd ACM Symp. Theory of Computing, 1971.

[5] M.R.GAREY, D.S.JOHNSON: *The complexity of near-optimal graph coloring.* J. of ACM, 23-1, 1976.

[6] J.HARTMANIS, L.BERMAN: *On isomorphisms and density of NP and other complete sets.* Proc. 8th ACM Symp. Theory of Computing, 1976.

[7] D.JOHNSON: *Approximation algorithms for combinatorial problems.* J. of Comp. Syst. Sci, 9,1974.

[8] R.M.KARP: *Reducibility among combinatorial problems.* In complexity of computer computations. R.E.Miller and J.W.Thatcher Eds. Plenum Prees, New York, 1972.

[9] S.MORAN, A.PAZ: *NP-optimization problems and their approximation.* Proc. 4th Int. Coll. on Automata, Languages and Programming, Turku, 1977.

[10] S.SAHNI: *Algorithms for scheduling independent tasks.* J. of ACM, 23-1, 1976.

[11] S.SAHNI, T.GONZALES: *P-Complete problems and approximate solutions.* Proc. IEEE 15th Symp. on Switching and Automata Theory, 1974.

[12] J.SIMON: *On Some Central Problems in Computational Complexity.* Cornell University TR 75-224, 1975.

A RECURSIVE APPROACH TO THE IMPLEMENTATION OF ENUMERATIVE METHODS

J.K. LENSTRA
Mathematisch Centrum, Amsterdam

A.H.G. RINNOOY KAN
Erasmus University, Rotterdam

ABSTRACT

Algorithms for generating permutations by means of both lexicographic and minimum-change methods are presented. A recursive approach to their implementation leads to transparent procedures that are easily proved correct; moreover, they turn out to be no less efficient than previous iterative generators. Some applications of explicit enumeration to combinatorial optimization problems, exploiting the minimum-change property, are indicated. Finally, a recursive approach to implicit enumeration is discussed.

KEY WORDS & PHRASES: *enumerative methods, recursive implementation, generation of permutations, lexicographic generator, minimum-change generator, combinatorial optimization, explicit enumeration, implicit enumeration, branch-and-bound.*

1. INTRODUCTION

The analysis of the inherent computational complexity of combinatorial problems indicates that for many of those problems a polynomial-time algorithm is not likely to exist. It appears that with respect to these problems we have to settle for some form of *enumeration* of the solution set whereby the feasible solutions are identified and an optimal one is obtained. For all but the smallest problems the number of feasible solutions is so large that the use of a computer for the actual computations is unavoidable. Thus, the computational performance of any enumerative method not only depends on algorithmic details but also on the computer implementation. The latter topic forms the subject of this paper.

More specifically, the paper will be devoted to a discussion of a *recursive* approach to the implementation of enumerative methods. We hope to demonstrate that such an approach leads to procedures that are elegant, easy to understand, easily programmed and easily proved correct. While these positive aspects will probably be recognized by most programmers, a familiar argument against recursive procedures suggests that none the less they require inordinate running times. Thus, ironically, many recursive approaches advocated in the literature are implemented after complicated manipulations in an iterative fashion [Barth 1968; Bitner *et al.* 1976; Gries 1975]! We will demonstrate on a simple example that with respect to efficiency a recursive implementation need certainly not be inferior to an iterative one; this remains true even if we consider a measure of efficiency that is computer and compiler independent.

The example referred to above is closely related to many combinatorial problems and involves the *generation of all permutations of a finite set*. In Section 2 we discuss various types of recursive permutation generators and present some results concerning their efficiency relative to iterative generators.

Since feasible solutions of many combinatorial problems are characterized by permutations, generators of permutations can be used in a straightforward way to solve such problems by *explicit enumeration* of all feasible solutions. We give some examples in Section 3, but it should be clear that this approach will solve only relatively small problems.

However, the advantages of a recursive approach carry through to forms of *implicit enumeration* as well. We illustrate this in Section 4 by presenting general frameworks for a popular type of implicit enumeration method known as *branch-and-bound*, in which again recursion plays a crucial role.

The material presented in this paper is adapted from [Lenstra & Rinnooy Kan 1975; Lenstra 1977].

2. GENERATION OF PERMUTATIONS

Methods for generating combinatorial configurations can often be classified as either *lexicographic* or *minimum-change* methods. The first mentioned type of method generates the configurations in a "dictionary" order, whereas the second type produces a sequence in which successive configurations differ as little as possible. The relative advantages of minimum-change methods have been discussed in the literature: the entire sequence is generated efficiently, each configuration being derived from its predecessor by a simple change; moreover, a minimum-change generator "may permit the value of the current arrangement to be obtained by a small correction to the immediate previous value" [Ord-Smith 1971].

The very "cleanliness" [Lehmer 1964] of these combinatorial methods allows a proper demonstration of what we believe to be the advantages of a recursive approach to the implementation of enumerative methods.

The algorithms which are to be presented in this section are defined as ALGOL 60 procedures. They contain no labels and generate the entire sequence of configurations after one call. Each time a new configuration has been obtained, a call of a procedure "problem" is made. Parameters of this procedure are the configuration and, for minimum-change algorithms, the positions in which it differs from its predecessor. The actual procedure corresponding to "problem" has to be defined by the user to handle each configuration in the desired way.

Previously published iterative generators usually have been organized in such a way that each call generates one configuration from its predecessor only. This necessitates continual recomputation of the information that is needed to find the next configuration in the sequence. A mechanism for performing this kind of computations efficiently has been described in [Ehrlich 1973A]. We do feel, however, that much of the clarity of essentially recursive algorithms is lost within any iterative implemenation.

For generators of various types of combinatorial configurations such as subsets, combinations and permutations, we refer to [Wells 1971; Ehrlich 1973A; Even 1973; Lenstra & Rinnooy Kan 1975; Reingold *et al.* 1977]. Permutation generators have been surveyed in [Lehmer 1964; Ord-Smith 1970, 1971; Sedgewick 1977].

In Section 2.1 a minimum-change generator of permutations is presented. It produces a sequence in which each permutation is derived from its predecessor by *transposing two adjacent elements*. Its basic principles have been discovered by Steinhaus [Gardner 1974] and were rediscovered independently in [Trotter 1962] and [Johnson 1963]. Trotter's iterative algorithm was for a number of years the fastest permutation generator. A more efficient iterative implementation has been presented in [Ehrlich 1973B]; see also [Gries 1975; Dershowitz 1975].

The lexicographic generator of permutations in Section 2.2 produces all permutations π of the set $\{1,\ldots,n\}$ in such a way that $\pi(n)\pi(n-1)\ldots\pi(1)$ is an *increasing n-ary number*.

In Section 2.3 our recursive generators are compared to previously published procedures.

2.1. A minimum-change generator

Given a set $\{\pi^*(1),\ldots,\pi^*(n)\}$, we define an undirected graph $G(n)$ whose vertices are given by the $n!$ n-permutations of this set; (π,ρ) is an edge of G(n) if and only if π and ρ differ only in two neighboring components. A hamiltonian path in G(n) corresponds to a sequence of permutations in which each permutation is derived from its predecessor by transposing two adjacent elements.

According to Steinhaus's method, we may construct such a sequence inductively as follows. For n = 1, it consists of the 1-permutation. Let the sequence of (n-1)-permutations be given. Placing $\pi^*(n)$ at the right of the first (n-1)-permutation, we obtain the first n-permutation. The n-1 next ones are obtained by successively interchanging $\pi^*(n)$ with its left-hand neighbor. After that, $\pi^*(n)$ is found at the left of the first (n-1)-permutation. Replacing this (n-1)-permutation by its successor in the (n-1)-sequence gives us the (n+1)-st n-permutation, and the n-1 next ones arise from successive transpositions of $\pi^*(n)$ with its right-hand neighbor. Then $\pi^*(n)$ is found at the right of the second (n-1)-permutation, which is now replaced by the third one, and the process starts all over again. It is easily seen that the first and last permutations in the sequence are given by $\pi^* = (\pi^*(1),\ldots,\pi^*(n))$ and $\rho^* = (\pi^*(2),\pi^*(1),\pi^*(3),\ldots,\pi^*(n))$ respectively; they are adjacent and thus we have found a hamiltonian circuit in G(n).

Steinhaus's method can be described more formally by a sequence S(2) of n!-1 transpositions. Denoting the transposition of $\pi^*(i)$ and the h-th element in the current permutation of $\{\pi^*(1),\ldots,\pi^*(i-1)\}$ by $i \leftrightarrow h$, we define the transposition sequence S(i) recursively by

$$S(i) = S(i+1), i \leftrightarrow h_1, S(i+1), i \leftrightarrow h_2, \ldots, S(i+1), i \leftrightarrow h_{i-1}, S(i+1)$$

where

$$h_k = \begin{cases} k & \text{if } \pi^*(i) \text{ moves rightwards,} \\ i-k & \text{if } \pi^*(i) \text{ moves leftwards,} \end{cases}$$

and S(n+1) is empty. Figure 1 and Table 1(mc) show the graphs G(n) for $n \le 4$ and the sequence for n = 4. Note that G(4) is the edge graph of a solid truncated octahedron, replicas of which fill entire 3-space. Similar statements of this remarkable property hold for all n [Lenstra Jr. 1973].

The following *minimum-change generator of permutations* produces the sequence described above.

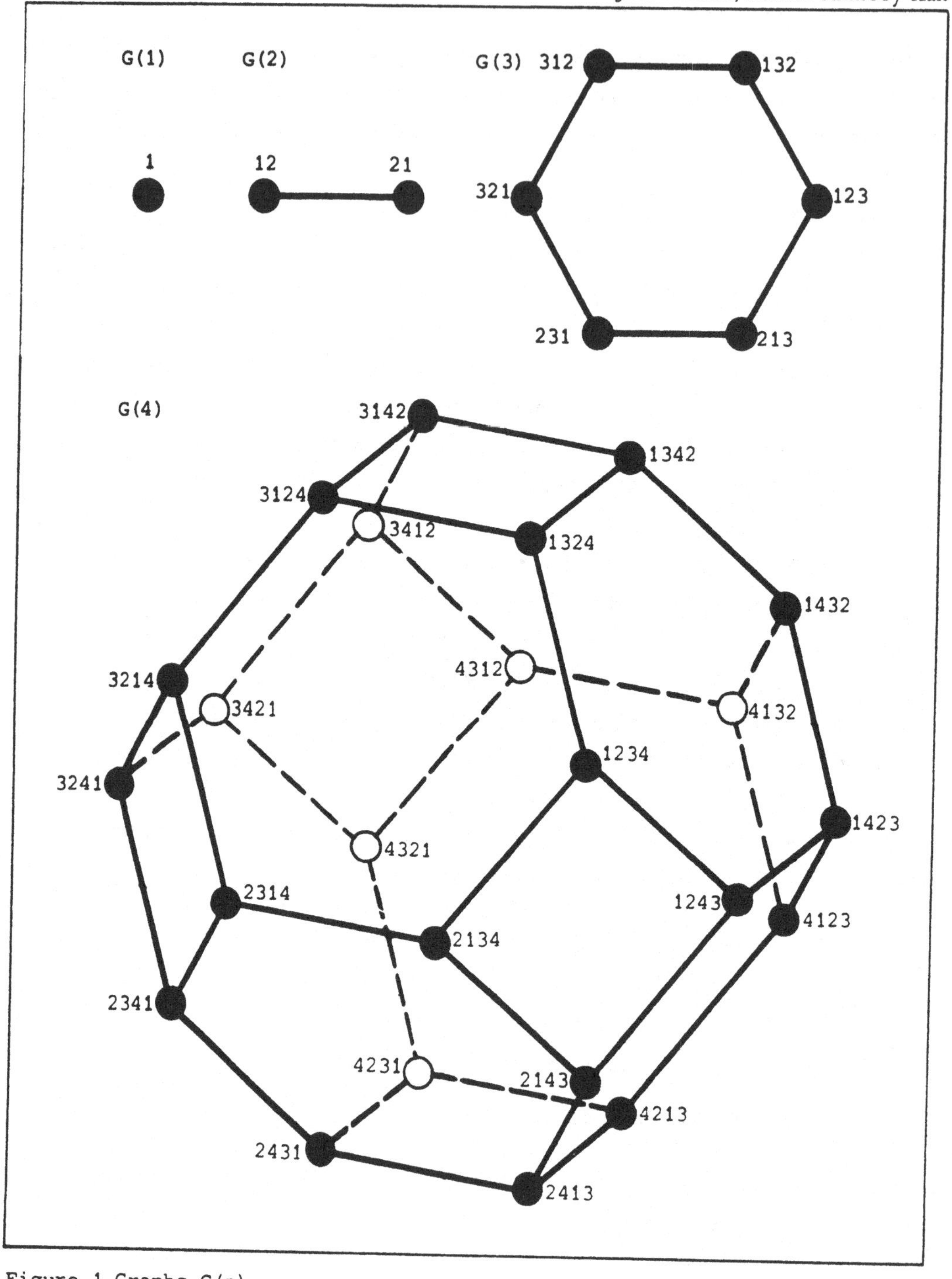

Figure 1 Graphs G(n).

TABLE 1. PERMUTATION SEQUENCES

	mc	lex
1	1234	4321
2	1243	3421
3	1423	4231
4	4123	2431
5	4132	3241
6	1432	2341
7	1342	4312
8	1324	3412
9	3124	4132
10	3142	1432
11	3412	3142
12	4312	1342
13	4321	4213
14	3421	2413
15	3241	4123
16	3214	1423
17	2314	2143
18	2341	1243
19	2431	3214
20	4231	2314
21	4213	3124
22	2413	1324
23	2143	2134
24	2134	1234

```
procedure pm mc (problem,n,pi); value n,pi;
integer n; array pi; procedure problem;
begin   real pin; integer k,q; boolean array r[1:n];

        procedure rite(i); value i; integer i;
        if i < n then
        begin   boolean rj; real pii; integer ti,j;
                pii:= pi[q]; j:= i+1;
                q:= q-1;
                rj:= r[j]; if rj then rite(j) else left(j);
                for ti:= 2 step 1 until i do
                begin   k: = q + ti;
                        pi[k-1]:= pi[k]; pi[k]:= pii; problem(pi,k-1);
                        rj:= ¬rj; if rj then rite(j) else left(j)
                end;
```

```
                r[j]:= ¬rj
        end     else
        begin   q:= 0;
                for k:= 2 step 1 until n do
                begin   pi[k-1]:= pi[k]; pi[k]:= pin; problem(pi,k-1)
                end
        end;

        procedure left(i); value i; integer i;
        if i < n then
        begin   boolean rj; real pii; integer ti,j;
                pii:= pi[q+i]; j:= i+1;
                rj:= r[j]; if rj then rite(j) else left(j);
                for ti:= i-1 step -1 until 1 do
                begin   k:= q+ti;
                        pi[k+1]:= pi[k]; pi[k]:= pii; problem(pi,k);
                        rj:= ¬rj; if rj then rite(j) else left(j)
                end;
                r[j]:= ¬rj;
                q:= q+1
        end     else
        begin   for k:= n-1 step -1 until 1 do
                begin   pi[k+1]:= pi[k]; pi[k]:= pin; problem(pi,k)
                end;
                q:= 1
        end;

        pin:= pi[n]; q:= 0; for k:= 2 step 1 until n do r[k]:= false;
        problem(pi,0); if n > 1 then left(2)
end pm mc.
```

A call "pm mc (problem,n,π^*)" has the following effect:
if n = 1, then a call "problem(π^*,0)" is made; else

- a hamiltonian path in G(n) from π^* to $\rho^* = (\pi^*(2),\pi^*(1),\pi^*(3),\ldots,\pi^*(n))$ is traversed;

- in vertex π^* a call "problem(π^*,0)" is made;
- in each vertex π, reached by transposition of the elements in position: k and k+1, a call "problem(π,k)" is made.

The latter two assertions are clear from inspection. To prove the first one, we note that a call "rite(i)" ("left(i)") performs a series of i-1 transpositions of $\pi^*(i)$ with its right (left) neighbor, where the predicate r(i) indicates which direction has to be chosen. By induction on i we can show that a call "rite(i)" or "left(i)" generates all permutations in which the current order of $\pi^*(1),\ldots,\pi^*(i-1)$ is preserved, only transposing adjacent elements, whereas just before such a call and immediately after its execution, π and q have the following property:

the indices $(i,\ldots,n)$ can be rearranged as $(j_1,\ldots,j_q,j_{q+i},\ldots,j_n)$ with $j_1 > \ldots > j_q$, $j_{q+i} < \ldots < j_n$, such that $\pi(k) = \pi^*(j_k)$ for $k = 1,\ldots,q,q+i,\ldots,n$.

The first assertion now corresponds to the effect of a call "left(2)", which indeed activates the whole process. This completes the proof.

Using the integer q to determine the place of the transpositions is easier and more efficient than keeping track of the inverse permutation for that purpose, as is done in [Ehrlich 1973A; Ehrlich 1973B].

In order to add to the transparency and efficiency of the procedure, two simple constructions have been applied. First, we have distinguished between the leftward and rightward moves of the elements by means of two procedures calling themselves and one another. Further, the deepest level of the recursion has been written out explicitly. This device clearly reduces the number of checks to see if the bottom of the recursion has been reached already; it enables us also to deal separately with the n-th element, which is involved in (n-1)/n of the transpositions.

We make one final remark on minimum-change sequences of permutations. Given an undirected graph H(n) on n vertices, we define an undirected graph $G_H(n)$ on the set of n-permutations; (π,ρ) is an edge of $G_H(n)$ if and only if π can be obtained from ρ by a single transposition of the elements in positions k and ℓ, where (k,ℓ) is an edge of H(n). One [Lenstra Jr. 1973] can prove that $G_H(n)$ *contains a hamiltonian circuit if and only if* H(n) *contains a spanning tree*. The "only if"-part is obvious; the "if"-part follows by an inductive argument. Note that the *transposition graph* H(n) of Steinhaus's method is a tree with edge set $\{(k,k+1) \mid k = 1,\ldots,n-1\}$.

2.2. A lexicographic generator

As mentioned before, the permutations π of the set $\{1,\ldots,n\}$ are ordered lexicographically when $\pi(n)\pi(n-1)\ldots\pi(1)$ is an increasing n-ary number. Table 1(lex) shows the sequence for n = 4.

Our *lexicographic generator of permutations* is given below. At each level of the recursion exactly one component of π is defined, and at the bottom a call "problem(π)" is made.

```
procedure pm lex (problem,n); value n;
integer n; procedure problem;
begin   integer h; integer array pi[1:n];

        procedure node(n); value n; integer n;
        if n = 1 then problem(pi) else
        begin   integer k,m,pin;
                m:= n-1; pin:= pi[n];
                node(m);
                for k:= m step -1 until 1 do
                begin   pi[n]:= h:= pi[k]; pi[k]:= pin; pin:= h;
                        node(m)
                end;
                for k:= n step -1 until 2 do pi[k]:= pi[k-1]; pi[1]:= pin
        end;

        for h:= n step -1 until 1 do pi[h]:= n+1-h;
        node(n)
end pm lex.
```

A call "pm lex (problem,n)" has the following effect:

- all permutations π of $\{1,\ldots,n\}$ are generated in lexicographic order;
- for each permutation π a call "problem(π)" is made.

To prove the first assertion, let us assume that, given a permutation π, a call "node(ℓ)" is made. It is easily checked that just before the ℓ calls "node(ℓ-1)" on the next level of the recursion, the then current permutation ρ is given by

$$\rho = (\rho(1),\ldots,\rho(k-1),\rho(k),\rho(k+1),\ldots,\rho(\ell-1),\rho(\ell),\rho(\ell+1),\ldots,\rho(n))$$
$$= (\pi(1),\ldots,\pi(k-1),\pi(k+1),\pi(k+2),\ldots,\pi(\ell),\pi(k),\pi(\ell+1),\ldots,\pi(n)),$$

for $k = \ell,\ell-1,\ldots,1$. By induction on ℓ it can be shown that a call "node(ℓ)" generates all permutations π in which $\pi(\ell+1),\ldots,\pi(n)$ remain unchanged, in increasing order, whereas just before such a call and immediately after its execution, π satisfies $\pi(1) > \pi(2) > \ldots > \pi(\ell)$. The observation that the effect of a call "node(n)" corresponds to the first assertion completes the proof.

2.3. Computational experience

The algorithms presented in Sections 2.1 and 2.2 have been compared to ALGOL 60 versions of two minimum-change generators, mentioned in the introduction to Section 2.

Table 2 shows the result of the comparison. The running times have been measured during one uninterrupted run on the Electrologica X8 computer of the Mathematisch Centrum; a procedure with an empty body was chosen for the actual parameter "problem". Our minimum-change algorithm turns out to be faster than corresponding previously published procedures. Although the time differences are not spectacular, a recursive approach should certainly not be rejected on grounds of computational inefficiency *a priori*.

Results like the above ones unavoidably remain computer and compiler dependent. It is of interest to note in this context that some experiments using PASCAL on the Control Data Cyber 73-28 of the SARA Computing Centre in Amsterdam instead of ALGOL 60 on the Electrologica X8 showed a nineteen-fold increase in speed for a recursive subset generator and a fourteen-fold increase for an iterative one. On the other hand, the running times of the iterative generators may be reduced by up to twenty percent by a different transformation of these generators into PASCAL procedures producing all configurations at one call.

In order to develop a computer independent measure of efficiency, let us define

$$a = \lim_{n\to\infty} \frac{\text{number of array subscript evaluations}}{\text{number of generated configurations}},$$

array access being a dominant factor in this type of ALGOL 60 procedure [Ord-Smith 1971]. For recursive algorithms, evaluation of a is accomplished by the solution of recursive expressions. For Trotter's iterative algorithm only a lower bound can be given; it is not clear if a finite limit exists.

TABLE 2. COMPARISON OF FOUR PERMUTATION GENERATORS

generator	restrictions	time	a
[Trotter 1962; Ord-Smith 1971]	$n \geq 2$	91.3	≥7
[Ehrlich 1973B]	$n \geq 3$, $n \neq 4$	58.1	3
pm mc	$n \geq 1$	42.9	3
pm lex	$n \geq 1$	92.4	6.44

time : CPU seconds on an Electrologica X8 for $n = 8$.

a : average array access (in the limit).

3. EXPLICIT ENUMERATION

Generators of combinatorial configurations can be used to solve many combinatorial optimization problems through enumeration and evaluation of all feasible solutions. Needless to say, only very small problems can be solved by such a brute force approach, even if the minimum-change property of the generators is exploited. However, they can be applied to validate more complicated solution methods by checking their results on small problems.

An an illustration we will show how generators of permutations can be used to solve sequencing problems P of the form

$$\min_\pi \{f_P(\pi)\}$$

where π runs over all permutations of $\{1,\ldots,n\}$. This formulation includes several well-known combinatorial optimization problems. Recall that the criterion function of the *quadratic assignment problem* (QAP) is given by

$$f_{QAP}(\pi) = \sum_{i=1}^n \sum_{j=1}^n c_{\pi(i)\pi(j)} d_{ij}$$

where (c_{ij}) and (d_{ij}) are nonnegative $n \times n$-matrices. If we take $d_{ij} = 1$ for $i > j$, $d_{ij} = 0$ otherwise, we obtain the *acyclic subgraph problem* (ASP).

Analogously, the choice $d_{12} = d_{23} = \ldots = d_{n-1,n} = d_{n1} = 1$, $d_{ij} = 0$ otherwise, leads to the *traveling salesman problem* (TSP), that is called *symmetric* if $c_{ij} = c_{ji}$ for all i,j.

If we define the *reflection* of π by $\bar{\pi} = (\pi(n),\ldots,\pi(1))$, it is obvious that $f_{ASP}(\bar{\pi}) = \sum_{i\neq j} c_{ij} - f_{ASP}(\pi)$ for the ASP and $f_{TSP}(\bar{\pi}) = f_{TSP}(\pi)$ for the symmetric TSP. It follows that for these two problems it suffices to enumerate a *reflection-free* set of permutations. Further, since $f_{TSP}((\pi(k+1),\ldots,\pi(n),\pi(1),\ldots,\pi(k))) = f_{TSP}(\pi)$ for any k, we may fix one of the components of π when solving a TSP. The $(n-1)!/2$ solutions to a symmetric TSP are the hamiltonian circuits in a complete undirected graph; they are called *rosary permutations* [Harada 1971; Read 1972; Roy 1973].

In the minimum-change generator of permutations, discussed in Section 2.1, the elements $\pi^*(1)$ and $\pi^*(2)$ are transposed half-way. If a permutation π is generated before this transposition, then its reflection $\bar{\pi}$ occurs thereafter. Hence the first $n!/2$ permutations form a relection-free set (*cf.* [Roy 1973]). Generally, the $n!/m!$ permutations preserving the original order of $\pi^*(1),\ldots,\pi^*(m)$ can be generated by a simple adaptation of "pm mc":

```
procedure pp mc (problem,n,m,pi); ...;
begin    ...
         ...; if n > m then left(m+1)
end pp mc1.
```

The above sequencing problems may now be solved by calls

pm mc (qap,n,π),

pp mc (asp,n,2,π),

pp mc (tsp,n-1,if symmetric then 2 else 1,π),

where "qap", "asp" and "tsp" are procedures which compute the changes occurring in the criterion functions of these problems.

4. IMPLICIT ENUMERATION

The permutation generator presented in Section 2.2 can easily be adapted to be used for implicit enumeration purposes by adding a lower bound calculation on all possible completions of a partial configuration. In the early fifties, Lehmer used such an approach to solve the linear assignment problem (!) [Tompkins 1956]. The fact that our recursive generators coupled with a simple lower bound may well outperform sophisticated implicit enumeration algorithms that suffer from a large computational overhead [Rinnooy Kan *et al.* 1975] underlines the applicability of recursive programming to implicit enumeration methods of the *branch-and-bound* type in general.

In this section we present a quasi-ALGOL description of branch-and-bound procedures, indicating in which case a recursive approach might be suitable. For a formal characterization of branch-and-bound procedures, we refer to the axiomatic framework in [Mitten 1970] and its correction in [Rinnooy Kan 1976]; see also [Agin 1966; Balas 1968] for analyses of the case in which the set of feasible solutions is finite and [Kohler & Steiglitz 1974] for the case of permutation problems. Some standard examples of branch-and-bound methods have been surveyed in [Lawler & Wood 1966].

Suppose then, that a *set* X *of feasible solutions* and a *criterion function* $f: X \rightarrow \mathbb{R}$ are given, and define the *set* X^* *of optimal solutions* by

$$X^* = \{x^* | x^* \in X,\ f(x^*) = \min\{f(x) | x \in X\}\}.$$

A branch-and-bound procedure to find an element of X^* can be characterized as follows.

- Throughout the execution of the procedure, the *best solution* x^* *found so far* provides an *upper bound* $f(x^*)$ on the value of the optimal solution.
- A *branching rule* b associates to $Y \subset X$ a family $b(Y)$ of subsets such that $\cup_{Y' \in b(Y)} Y' \cap X^* = Y \cap X^*$; the subsets Y' are the *descendants* of the *parent* subset Y. This rule only has to be defined on a class $\mathcal{X}$ with $X \in \mathcal{X}$ and $b(Y) \subset \mathcal{X}$ for any $Y \in \mathcal{X}$.
- A *bounding rule* $lb: \mathcal{X} \rightarrow \mathbb{R}$ provides a *lower bound* $lb(Y) \leq f(x)$ for all $x \in Y \in \mathcal{X}$. *Elimination* of Y occurs if $lb(Y) \geq f(x^*)$.
- A *predicate* $\xi: \mathcal{X} \rightarrow \{\underline{true}, \underline{false}\}$ indicates if during the examination of Y (*e.g.* during the calculation of $lb(Y)$) a feasible solution $x(Y)$ is generated which has to be evaluated. *Improvement* of x^* occurs if $f(x^*) > f(x(Y))$.

- A *search strategy* chooses a subset from the collection of generated subsets which have so far neither been eliminated nor led to branching.

It turns out that, of the three search disciplines that have been used most frequently, two are suitable for recursive implementation. To illustrate this point, we shall now present three general procedures:

- "bb jumptrack" implements a *breadth-first search* where a subset with minimal lower bound is selected for examination; this type of tree search is known as *frontier search*;
- "bb backtrack1" implements a *depth-first search* where the descendants of a parent subset are examined in an arbitrary order; this type is known as *newest active node search*;
- "bb backtrack2" implements a *depth-first search* where the descendants are chosen in order of nondecreasing lower bounds; this type is sometimes called *restricted flooding*.

During the tree search, the parameters na and nb count the numbers of subsets that are eliminated and that lead to branching respectively. We define the operation ":zϵ" in the statement "s:zϵ S" to mean that s:= s^* with $z(s^*) = \min\{z(s)|s \in S\}$; hence, ":$\epsilon$" indicates an arbitrary choice.

```
procedure bb jumptrack (X,f,x*,b,lb,ξ,na,nb);
begin   local Y,Y',B ⊂ X, Y,Y' ∈ X, LB: X → ℝ;
        na:= nb:= 0; Y:= ∅;
        LB(X):= lb(X); if ξ(X) then x*:f∈ {x*,x(X)};
        if LB(X) ≥ f(x*) then na:= 1 else Y:= {X};
        while Y ≠ ∅ do
        begin   Y:LB∈ Y;
                nb:= nb+1; B:= b(Y); Y:= (Y-{Y})∪B;
                while B ≠ ∅ do
                begin   Y':∈ B; B:= B-{Y'};
                        LB(Y'):= lb(Y'); if ξ(Y') then x*:f∈ {x*,x(Y')}
                end;
                Y':= {Y'|Y' ∈ Y, LB(Y') ≥ f(x*)};
                na:= na+|Y'|; Y:= Y-Y'
        end
end bb jumptrack.
```

```
procedure bb backtrack1 (X,f,x*,B,lb,ξ,na,nb);
begin   local Y' ∈ X;

        procedure node(Y);
        begin   local B ⊂ X, LB ∈ ℝ;
                LB:= lb(Y); if ξ(Y) then x*:f∈ {x*,x(Y)};
                if LB ≥ f(x*) then na:= na+1 else
                begin   nb:= nb+1; B:= B(Y);
                        while B ≠ ∅ do
                        begin   Y':∈ B; B:= B-{Y'};
                                if LB < f(x*) then node(Y')
                        end
                end
        end;

        na:= nb:= 0;
        node(X)
end bb backtrack1.

procedure bb backtrack2 (X,f,x*,B,lb,ξ,na,nb);
begin   local B ⊂ X, Y' ∈ X, LB: X → ℝ;

        procedure node(Y);
        begin   local Y ⊂ X;
                nb:= nb+1; Y:= B:= B(Y);
                while B ≠ ∅ do
                begin   Y':∈ B; B:= B-{Y'};
                        LB(Y'):= lb(Y'); if ξ(Y') then x*:f∈ {x*,x(Y')}
                end;
                while Y ≠ ∅ do
                begin   Y':LB∈ Y; Y:= Y-{Y'};
                        if LB(Y') ≥ f(x*) then na:= na+1 else node(Y')
                end
        end;
```

```
        na:= nb:= 0;
        LB(X):= lb(X); if ξ(X) then x*:f∈ {x*,x(X)};
        if LB(X) ≥ f(x*) then na:= 1 else node(X)
end bb backtrack2.
```

Anyone familiar with branch-and-bound will have noticed that the above descriptions provide only a minimal algorithmic framework. Numerous problem-dependent variations may be included in an actual procedure. For instance, elimination of Y may be possible already during the calculation of lb(Y) or may be due to *elimination criteria* based on dominance rules or feasibility considerations. In a minor (and in our experience quite successful) variation on "bb backtrack1", the descendants Y' of a parent subset Y are not chosen arbitrarily, but according to some heuristic, *e.g.* preliminary lower bounds lb'(Y') with lb(Y) ≤ lb'(Y') ≤ lb(Y'). Many similar variations are possible but do not affect the basic mechanisms outlined above.

A main characteristic of many branch-and-bound procedures is the unpredictability of their computational behavior. Their worst-case performance may be close to explicit enumeration, and no satisfying analyses of average-case behavior have been presented up to now [Karp 1976; Lenstra & Rinnooy Kan 1978]. Extensive computational experience seems to be the only way to test their quality. Branch-and-bound should not be used before one feels sure that the complexity of the problem is such that no better approach can be found. However, this is often the case, and methods of branch-and-bound are widely used for solving combinatorial optimization problems.

From our experience with the implementation of branch-and-bound algorithms we may conclude that again the recursive approach produces transparent procedures, in which much administrative work is taken over by the compiler without a noticeable negative effect on overall efficiency.

ACKNOWLEDGMENTS

The authors gratefully acknowledge the valuable help and suggestions from J.D. Alanen, P. van Emde Boas, J.S. Folkers, B.J. Lageweg, H.W. Lenstra, Jr., G.K. Manacher and I. Pohl. This research was partially supported by NATO Special Research Grant 9.2.02 (SRG.7).

REFERENCES

N. AGIN (1966) Optimum seeking with branch-and-bound. *Management Sci.* 13, B176-185.

E. BALAS (1968) A note on the branch-and-bound principle. *Operations Res.* 16,442-445,886.

W. BARTH (1968) Ein ALGOL 60 Programm zur Lösung des traveling Salesman Problems. *Ablauf- und Planungsforschung* 9,99-105.

J.R. BITNER, G. EHRLICH, E.M. REINGOLD (1976) Efficient generation of the binary reflected Gray code and its applications. *Comm. ACM* 19,517-521.

N. DERSHOWITZ (1975) A simplified loop-free algorithm for generating permutations. *BIT* 15,158-164.

G. EHRLICH (1973A) Loopless algorithms for generating permutations, combinations and other combinatorial configurations. *J. Assoc. Comput. Mach.* 20,500-513.

G. EHRLICH (1973B) Algorithm 466, Four combinatorial algorithms. *Comm. ACM* 16,690-691.

S. EVEN (1973) *Algorithmic Combinatorics*, Macmillan, London.

M. GARDNER (1974) Some new and dramatic demonstrations of number theorems with playing cards. *Sci. Amer.* 231,122-125.

D. GRIES (1975) Recursion as a programming tool. Technical Report 234, Department of Computer Science, Cornell University, Ithaca.

K. HARADA (1971) Generation of rosary permutations expressed in hamiltonian circuits. *Comm. ACM* 14,373-379.

S.M. JOHNSON (1963) Generation of permutations by adjacent transposition. *Math. Comp.* 17,282-285.

R.M. KARP (1976) The probabilistic analysis of some combinatorial search algorithms. In: J.F. TRAUB (ed.) (1976) *Algorithms and Complexity: New Directions and Recent Results*, Academic Press, New York, 1-19.

W.H. KOHLER, K. STEIGLITZ (1974) Characterization and theoretical properties of branch-and-bound algorithms for permutation problems. *J. Assoc. Comput. Mach.* 21,140-156.

E.L. LAWLER, D.E. WOOD (1966) Branch-and-bound methods: a survey. *Operations Res.* 14,699-719.

D.H. LEHMER (1964) The machine tool of combinatorics. In: E.F. BECKENBACH (ed.) (1964) *Applied Combinatorial Mathematics*, Wiley, New York, 5-31.

H.W. LENSTRA, JR. (1973) Private communications.

J.K. LENSTRA (1977) *Sequencing by Enumerative Methods*, Mathematical Centre Tracts 69, Mathematisch Centrum, Amsterdam.

J.K. LENSTRA, A.H.G. RINNOOY KAN (1975) A recursive approach to the generation of combinatorial configurations. Report BW50, Mathematisch Centrum, Amsterdam.

J.K. LENSTRA, A.H.G. RINNOOY KAN (1978) On the expected performance of branch-and-bound algorithms. *Operations Res.* 26,347-349.

L.G. MITTEN (1970) Branch-and-bound methods: general formulation and properties. *Operations Res.* 18,24-34.

R.J. ORD-SMITH (1970) Generation of permutation sequences: part 1. *Comput. J.* 13,152-155.

R.J. ORD-SMITH (1971) Generation of permutation sequences: part 2. *Comput. J.* 14,136-139.

R.C. READ (1972) A note on the generation of rosary permutations. *Comm. ACM* 15,775.

E.M. REINGOLD, J. NIEVERGELT, N. DEO (1977) *Combinatorial Algorithms: Theory and Practice*, Prentice-Hall, Englewood Cliffs, N.J.

A.H.G. RINNOOY KAN (1976) On Mitten's axioms for branch-and-bound. *Operations Res.* 24,1176-1178.

A.H.G. RINNOOY KAN, B.J. LAGEWEG, J.K. LENSTRA (1975) Minimizing total costs in one-machine scheduling. *Operations Res.* 23,908-927.

M.K. ROY (1973) Reflection-free permutations, rosary permutations, and adjacent transposition algorithms. *Comm. ACM* 16,312-313.

R. SEDGEWICK (1977) Permutation generation methods. *Comput. Surveys* 9, 137-164,314.

C. TOMPKINS (1956) Machine attacks on problems whose variables are permutations. *Proc. Sympos. Appl. Math.* 6, Amer. Math. Soc., Providence, 195-211.

H.F. TROTTER (1962) Algorithm 115, Perm. *Comm. ACM* 5,434-435.

M.B. WELLS (1971) *Elements of Combinatorial Computing*, Pergamon, Oxford.

DATA STRUCTURES FOR COMBINATORIAL PROBLEMS

FABRIZIO LUCCIO
University of Pisa

The concept of data structure is widely known. We will not attempt here to give a definition of data structure, nor to describe the semantics of the operations required by such structures. Nor we will sistematically present a way a data structure can be implemented. The reader is referred, for example, to [1] for a comprehensive presentation of the above matters. In this lecture instead, we will try to put into evidence the importance of selecting a proper data structure in the solution of a given problem, by discussing different aspects of a specific working example.

A point must always be clear, namely, all algorithms are usually designed to run in some computer system, where the central store has specific characteristics. Unless differently stated, such a store is assumed to be random-access and addressable. Memories with different characteristics will be considered in the next lecture, and some new problems arising in data structure and algorithm design will be discussed.

1. A WORKING EXAMPLE

To investigate some aspects of data structure design we consider the "Floating Currency Problem" and its implications. This problem has been proposed by Herbert Freeman through a witty fable, that sounds approximately as follows [2].

"The country of X has an ancient democratic tradition. All citizens invest their money in foreign currencies, and the richest citizen is made king.

For centuries the international exhange rates have been relatively steady, which has given X peace and stability. Recently, however, several countries have left their currencies to float, thus causing continuous changes in the dynasty. Worst of all, the National Bank of X (which is in charge of proclaiming the king by midnight) has to face formidable peaks of computation load every night at 23:45, when a messanger reaches X with the quotations of the day.

The problem is then: is it possible to preprocess the assets of all citizens, to ease the proclaimation work?"

The first point to consider is how to organize the data. We let

$$V = \{v_1, v_2, \ldots, v_n\}$$

be the set of citizens, each represented as a d-dimensional vector

$$v_i = (x_1(v_i), x_2(v_i), \ldots, x_d(v_i)).$$

$x_j(v_i)$ is the amount of corrency j owned by citizen v_i. We assume that all $x_j(v_i)$ are non negative integers, however, several such components are likely to be equal to zero, as each citizen may invest in a limited number of favorite currencies.

An example of assets for five differentcitizens ($n = 5$), and five different currencies ($d = 5$), is reported in figure 1.a. The information is arranged in *array* form, with obvious meaning. Since such an array has size $n \times d$, any nontrivial algorithm on it is likely to take a number of steps of order at least nd, which is the minimum time required to read the input. Such a proposition immediately derives from a known result on adjacency-array based algorithms for graphs [3].

An alternative form of representation is shown in figure 1.b. Each citizen v_i is followed by a list of all currencies owned by v_i, taken in any order. These lists will be called *adjacency lists*, after a similar well known form of representation for graphs [1]. The size of the structure is now proportional to the number of nonzero entries in the previous array. An algorithm that requires only this information (e.g. an algorithm that doubles the assets of all citizens) would generally be faster if working on the adjacency lists, instead of the array. A discussion on two similar forms of representation for graphs, and their influence on algorithms, can be found in [4].(1)

Finally, the size of the list structure can be further reduced if the lists are merged as much as possible, in such a way that the maximum number of common assets are reported

---

(1) Note incidentally that the information of our problem can be represented in a bipartite graph, where citizens and currencies constitute the two families of nodes.

	LIT	FF	\$	GM	DF
v_1	5	8	7	0	0
v_2	2	0	3	0	3
v_3	0	8	3	0	0
v_4	0	0	7	1	0
v_5	0	8	3	2	0

(a)

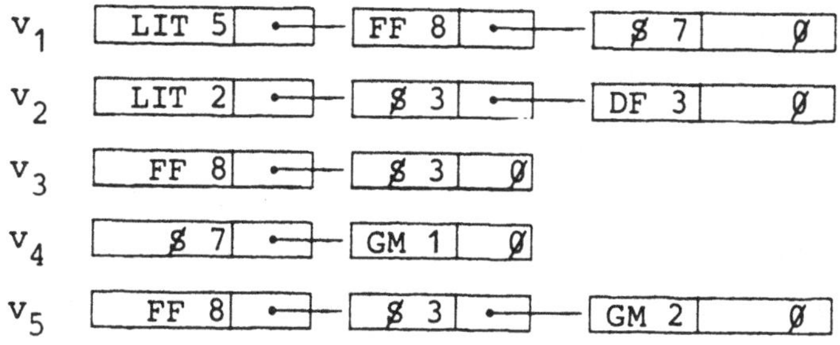

(b)

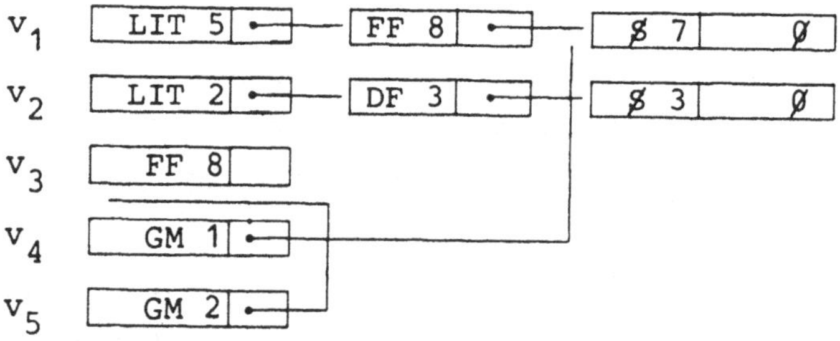

(c)

Fig. 1. The assets of five citizens represented in array form (a), adjacency lists (b), compact adajacency lists (c).

at the end of the lists, and are shared by different lists. This is indicated in the *compact adjacency lists* of figure 1.c. Such a contraction require reordering the list in a way that a global minimum is attained in the size of the structure. An algorithm working on the currency assets may its fastest version on compact adjacency lists.

Note that adjacency lists can be built from the array, and vice versa, in order nd steps. To build compact adajecency lists from one of the previous structures requires instead an exponential algorithm. By anticipating a concept that will be presented in a companion set of lectures, the problem of building compact adjacency lists is NP-complete. This can be seen as a simple variation of the "pruned trie space minimization" problem [5].

2. SEARCH, INSERT AND DELETE

Prior to the problem of king proclaimation, the one of updating the data of the population of X may be considered.

Search for a citizen v_i, or for a particular component $x_j(v_i)$, is required to update the assets of v_i after some financial operation. If no birth or death is observed in X over long periods of time, the structure of figure 1 remain unchanged, except possibly for the values of x_j's. Such structures are then said to be static, and search for v_i can be performed in constant time in a random-access memory, where the address of v_i is an easily computable function of i. If all the assets of v_i are to be changed, array form and adjacency lists are equivalent. If, however, access to a particular asset is required, only the array form allows accessing such an element in constant time.

If *insertions* (due to births) or *deletions* (due to deaths) of citizens are required in the structure, a dynamic organization must be employed, to allow efficient execution of such operations. This can be obtained via a binary balanced tree with nodes associated to the names of the citizens [6]. A node associated to v_i includes also a pointer to the assets of v_i, that is to the array row or adjacency list of v_i. Search, insert and delete in the tree can be executed in time proportional to log n. However, adding or deleting lists of assets is more efficiently done in the adjacency lists, than in the array form. All concepts supporting these points can be found in [1].

3. FINDING THE MAXIMA OF A SET OF VECTORS

Once the assets of all citizens are set up, and the quotations of all currencies are known, proclaiming the king is a matter of carrying out multiplications and additions. This computational work can be diminuished if it is limited to the richest citizens, thus excluding all the citizens which can be recognized to be poorer than some other citizen, independently of the rates of exchange of the various currencies. Preprocessing then consists in finding such richest citizens.

We now put the problem in precise terms. On the set $V = \{v_1, v_2, \ldots, v_n\}$ already introduced we define a partial ordering $\leq$, such that:

$$v_h \leq v_k \Leftrightarrow x_j(v_h) \leq x_j(v_k) \text{ , for all } j.$$

The problem is then the one of finding all the maximal elements of V under $\leq$. This problem has been introduced and thoroughly investigated in [7]. Successively a different approach has been independently developed in [8], yielding the same results of [7]. We report here the basic algorithm devised in [7], and then we specify such an algorithm for d = = 2,3 to show how a proper selection of data structures can increase its efficiency. For this purpose, let us introduce some notation.

p is a function from d-dimensional vectors to their (d-1)-dimensional projections on the second, ... , d-th component. That is:

$$p(v_i) \triangleq (x_2(v_i), x_3(v_i), \ldots, x_d(v_i)).$$

P_i , $0 \leq i \leq n$, are sets of (d-1)-dimensional vectors defined as:

$$P_0 \triangleq \emptyset,$$

$$P_i \triangleq \{p(v_j) : j \leq i\}, \quad 1 \leq i \leq n.$$

Finally, T_i is the set of maxima of P_i.

The algorithm to find all maxima of V is as follows. We make the assumption that $x_i(v_h) \neq x_i(v_k)$ for every i and $h \neq k$. (It can be easily proved that such an assumption makes the algorithm more easily understandable, but it does not alter its essential characteristics [7]).

1. Arrange the elements of V as a sequence $v_1, \ldots, v_n$ such that $x_1(v_i) > x_1(v_j)$ for $1 \leq i < j \leq n$. (I.e., sort on x_1 and rename the elements);

2. $i := 1$; $T_0 := \emptyset$;

3. while $i \leq n$ do

   begin

   if $\exists w \in T_{i-1} : p(v_i) < w$ then $T_i := T_{i-1}$

   else ($\underline{v}_i$ *is a maximum*)

construct T_i from P_i ;

i := i+1

end.

The validity of the algorithm can be readily established. We leave this proof to the reader.

From a computational point of view, the number of operations required by the algorithm is essentially due to steps 1 and 3. Step 1 require approximately n log n operations for sorting. Step 3 repeate n times a test of nontrivial implementation, namely $\exists w \in T_{i-1} : p(v_i) < w$; and can repeat up to n times the difficult operation: construct T_i from P_i. We now specify how to realize this step for $d = 2$ and $d = 3$.

d = 2. This is a very simple case. P_{i-1} is the set of the i-1 integers $x_2(v_j)$ for $1 \leq j \leq i-1$, whence T_{i-1} merely consists of the maximum w_{i-1} of such integers. The test in the if clause reduces to the single comparison between $p(v_i)=x_2(v_i)$ and w_{i-1}. Moreover, if $x_2(v_i) > w_{i-1}$, then T_i is immediately built as being composed of the single element $x_2(v_i)$. The total computational work of the algorithm is therefore of the order of n log n+n.

d = 3. P_{i-1} is a set of bidimensional vectors on x_1, x_2, and T_{i-1} is the set of maxima of such vectors, say $T_{i-1} = \{u_1,\ldots,u_s\}$. Without loss of generality we assume inductively that $x_2(u_h) > x_2(u_k)$ for $1 \leq h < k \leq s$. Since $u_1,\ldots,u_s$ are bidimensional maximal vectors, the above assumption implies that $x_3(u_h) < u_3(u_k)$ for $1 \leq h < k \leq s$. We also assume that $x_j(v_h) > 0$ for every j and h. Then, we rewrite step 3 of the algorithm in the following explicit form:

3a. Insert in T_0 the two dummy vectors $u'=(x_2(u')=\infty, x_3(u') = 0)$ and $u''=(x_2(u'')=0, x_3(u'')=\infty)$ (In the succeeding configurations of T_{i-1}, u' and u" will respectively constitute the new extreme vectors u_0 = u' and u_{s+1} = u");

3b. while $i \leq n$ do

begin

3b.1. determine the largest value j^+ of index j such that $x_2(v_i) < x_2(u_j)$, $u_j \in T_{i-1}$;

if $x_3(v_i) < x_3(u_{j^+})$ then $T_i := T_{i-1}$

else ($\underline{v}_i$ is a maximum)

begin

3b.2. determine the smallest value j^{++} of index j such that $x_3(v_i) < x_3(u_j)$, $j^+ < j \leq s+1$;

3b.3. construct T_i as $T_i := T_{i-1} - \{u_j : j^+ < j < j^{++}\} + \{p(v_i)\}$

end;

i:=i+1

end.

Before discussing the complexity of the new step 3, let us discuss the example of figure 2, where the vectors $p(v_i)$ are graphically represented as points in 2-dimensional space x_2, x_3. After executing step 3b for i = 1,2, the set T_2 shown in figure 2.b is easily found. Both v_1 and v_2 are maxima. When step 3b is repeated for i = 3, we have $x_2(v_3) < x_2(u_0)$, $x_2(u_1)$, that is $j^+ = 1$. Since $x_3(v_3) < x_3(u_1)$, v_3 is not a maximum.

When step 3b is repeated for i = 4, we have $j^+ = 0$, $x_3(v_4) > x_3(u_0)$ therefore v_4 is a maximum. We also have $x_3(v_4) < x_3(u_2)$, $x_3(v_4) < x_3(u_3)$, that is $j^{++} = 2$, therefore T_4 is built as $T_4 := T_3 - \{u_1\} + \{p(v_4)\}$.

From a computational point of view, the difficult ope-

	x_1	x_2	x_3
v_1	10	8	4
v_2	8	3	8
v_3	6	6	2
v_4	3	10	7
⋮			
v_n			

(a)

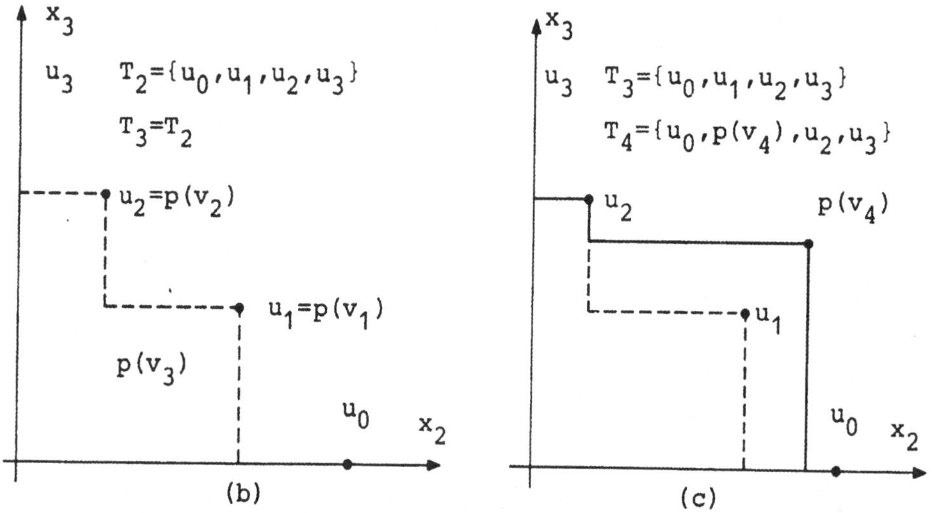

Figure 2. The functioning of step 3b. (a) Sample vectors in 3-dimensional space. (b) Comparison between $p(v_3)$ and the vectors of T_2. (c) Comparison between $p(v_4)$ and the vectors of T_3: v_4 is a maximum, and T_4 is consequently rebuilt.

rations of step 3 are the determination of j^+ (step 3b.1); the determination of j^{++} (step 3b.2); and the construction of T_i (step 3b.3). Here the use of a proper data structure is critical. We store the information contained in T_{i-1} in a balanced binary tree (see section 2), where $u_0,\dots,u_{s+1}$ are ordered according to the decreasing values of x_2. As alrrady nored, such elements are then automatically ordered for increasing values of x_3.

The determination of j^+ requires a search in the tree, that can be done in order log n operations. Since step 3b.1 is executed n times, in the algorithm, it contributes with a total cost of n log n operations.

The determination of j^{++}, and the consequent construction of T_i, is performed by searching in the tree all vectors u_j, $j = j^+ + 1,\dots,j^{++}$, and by removing such vectors from the tree, except for v_j++. Since each operation involves a work proportional to log n, and at most order of n vectors can be searched for and removed from the tree throughout the algorithm, then steps 3b.2 and 3b.3 contribute with a total work at most proportional to n log n. Therefore, the computational cost of the algorithm is of the order of n log n+n, as we already found for d = 2. Note once more that this result for d = 3 has been made possible by the use of very sophisticated data structures.

REFERENCES

[1] E.HOROWITZ and S.SAHNI: *Fundamentals of Data Structures.* Pitman, London 1977.

[2] H.FREEMAN: The floating currency problem. *Unpublished manuscript.*

[3] M.L.FISHER: Worst-Case Analysis of Heuristics. In "Interfaces Between Computer Science and Operations Research". *Proceedings of a Symposium held at the Mathematisch Centrum,* Amsterdam, September 7_10, 1976.

[4] P.VAN EMDE BOAS: Developments in Data Structures. In "Interfaces Between Computer Science and Operations Research". *Proceedings of a Symposium held at the Matematisch Centrum,* Amsterdam, September 7_10, 1976.

[5] M.R.GAREY and D.S.JOHNSON: *Computers and Intractability.* Freeman and Co., San Francisco 1979.

[6] D.E.KNUTH: *The Art of Computer Programming.* Vol. 3. Addison-Wesley, Reading 1973.

[7] H.T.KUNG, F.LUCCIO and F.P.PREPARATA: On Finding the Maxima of a Set of Vectors. *J. ACM* 22, Oct. 1975, 469-476.

[8] Y.A.KRIUKOV: Optimal Algorithms of Definition of Pareto Optimal Set. *Proc. Tenth International Symposium on Mathematical Programming,* Montreal 1979.

DATA STRUCTURES FOR BIDIMENSIONAL MEMORY

FABRIZIO LUCCIO
University of Pisa

The study of bidimensional data organization in a general sense has been initiated in 1976 [1]. Previously, the only contribution of some interest had been specifically referred to drum-type storage [2].

In bidimensional organization data are allocated in cells of a chessboard-type storage, where each cell c is denoted by two integer coordinates: $c = (x,y)$. Adjacency is in two dimensions, that is $c_1 = (x_1,y_1)$ and $c_2 = (x_2,y_2)$ are adjacent if $x_1 = x_2$ and $|y_1-y_2| = 1$, or $y_1 = y_2$ and $|x_1-x_2| = 1$. Aside from its theoretical interest, this organization is directly applicable in a new type of mass memory, the so called magnetic bubble lattice file [3].

In bidimensional organization a *figure* is defined as a connected set of cells. *Rectangles* are figures of particular relevance [4,5]: a basic problem is to decompose a figure F into m rectangles $R_1,\dots,R_m$, such that $F = \bigcup_{i=1}^{m} R_i$. In fact, the set $\{R_1, \ldots, R_m\}$ is called a *description* of F. A *minimal description* (MD) is one containing minimum number of rectangles. A *minimal disjoint description* (MDD) is an MD where all rectangles are pairwise disjoint.

The problems of finding an MD and an MDD for a given figure have been respectively treated in [4] and [5]. The solution proposed for the former problem require an exponential number of steps in the worst case. In fact, it has been successively proved that this problem is NP-complete [6]. The latter problem instead has been solved in [5] in polynomial time. It is worth to note that the same problem is one of the open problems proposed by Klee [7] in a quite different mathematical environment.

Studies of the above type have succeedingly become relevant in the fields of VLSI design [8], and geographic data processing [9]. These have also justified further studies to determine the contour of a figure F from $R_1, \ldots, R_m$ [10], and for solving path problems with rectilinear metric [11,12].

In this lecture we will review the basic properties of MD and MDD, and discuss the algorithms proposed for their determination.

1. FINDING MD

To find a minimal description (MD) of a figure F, we must examine some properties of the figures in our organization. First, a description $\{R_1, \ldots, R_m\}$ of F is *irredundant* if for any k, $1 \leq k \leq m$, we have $\bigcup_{i=1}^{m} R_{i\,(i \neq k)} \subset F$. All descriptions considered here are irredundant, without need of specification.

Let e denote the number of edges of F. It is straightforward to prove that the order of magnitude of e is $\Omega(m)$ and $O(m^2)$ (for the use of notation Ω and O see [13]).

A basic concept is the one of *prime rectangle* P, defined as a rectangle $P \subseteq F$, such that no rectangle P' exists with $P' \subseteq F$, $P \subset P'$. Intuitively prime rectangles are rectangular portions of F of maximal size. Their importance is substantiated by the easily provable fact that at least one MD is uniquely composed of prime rectangles. Hence, we can generate all prime rectangles and construct an MD as a selection of such rectangles.

A non trivial result proved in [4] shows that the number of prime rectangles is upper bounded by $\lfloor e^2/16 \rfloor$, and that this bound is strict. In figure 1 we show an example with $\lfloor e^2/16 \rfloor$ prime rectangles. An algorithm is given in [4] to find all prime rectangles of a figure $F = \bigcup_{i=1}^{m} R_i$ in $O(m^3)$ steps.

Finding an MD is an NP-complete problem [6]. Therefore we must try to develope an algorithm reasonably efficient in most cases. We can divide F in all the *elementary parts* obtained by intersecting all prime rectangles, and draw a covering table where each prime rectangle covers a set of elementary parts. Then, the problem of deriving MD is reduced to a classical covering problem.

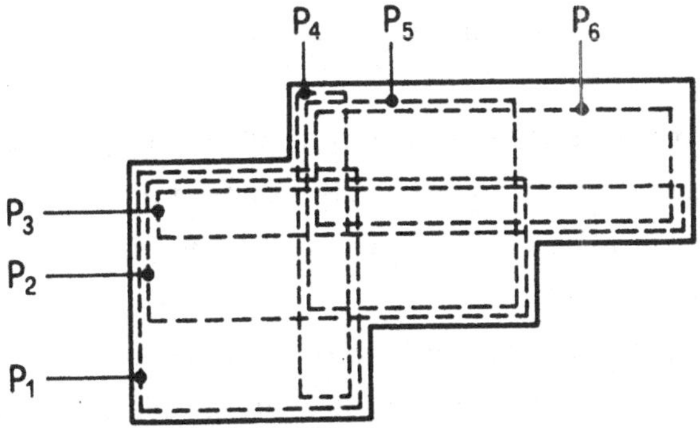

Fig. 1. A figure with e = 10, and $\lfloor e^2/16 \rfloor = 6$ prime rectangles P_1 to P_6.

It is straightforward to define a prime rectangle P as *essential*, if there exists an elementary part of F covered by the only prime rectangles P. Hence, all essential prime rectangles will be contained in the solution. An unexpected result proved in [4] shows that any figure has at least one essential prime rectangle. This result seems to suggest that an MD could be found by successive selections of essential rectangles, thus leading to a polynomial time algorithm. The apparent contradiction is eliminated by noting that, once an essential rectangle is selected as part of an MD, the residue portion of F does not necessarily contain an essential prime rectangle. The example of figure 2 illustrates this point.

2. FINDING MDD

The notion of prime rectangle looses its relevance in the study of descriptions where all rectangles are disjoint. A crucial role is instead played by the corners where the figure is concave (called *concave corners*). Other elements are also important. The *line* is defined as any straight line segment internal to a figure F (the end points of the line may lie on the contour of F). Thus, two edges of F are said to be *aligned* if they lie on the same straight line, and can be connected by a line.

The above entities are illustrated in figure 3 (note that F has a hole). C_1 to C_7 are the concave corners. g_1 to g_5 are instances of lines. e_1-e_2, e_3-e_4, e_5-e_6, are the couples of aligned edges.

Finally we define a *set of heavy lines* as a set of lines connecting couples of aligned edges, such that no two lines in the set are incident (i.e., have one point in common. Obviously only two orthogonal lines can be incident). A set of heavy lines S is *complete* i no other set of heavy lines S'

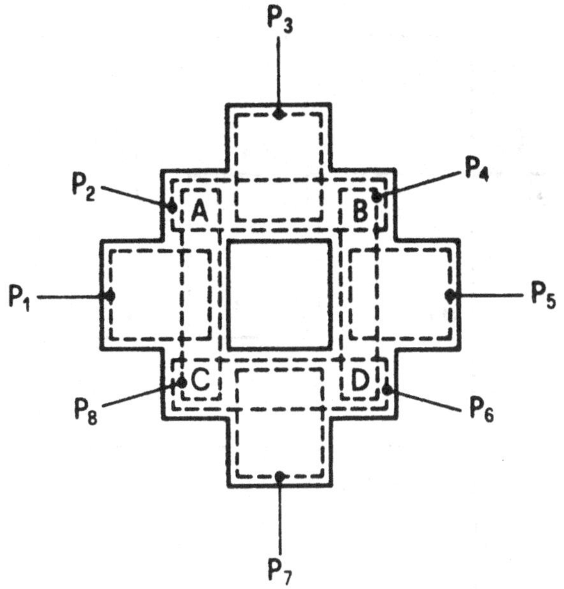

Fig. 2. Among the eight prime rectangles P_1 to P_8, rectangles P_1, P_3, P_5 and P_7 are essential 4. After elimination of essential rectangles, the residue ABCD does not contain any new essential rectangle.

exists, with $S' \supset S$. In figure 3 we have two complete sets of heavy lines: $\{g_1\}$ and $\{g_2, g_3\}$. (Note that g_1 and g_2 are incident, hence only one of these lines can be contained in a set. Similarly g_1 and g_3 are incident).

We can now illustrate the basic properties of a disjoint description. For each *figure without aligned edges*, it can be proved that the number of rectangles in an MDD is related to the number of concave corners and to the number of holes by the simple relation [5]:

$$\#MDD = \#(\text{concave corners}) - \#(\text{holes}) + 1.$$

Moreover, an MDD can be simply constructed by tracing one

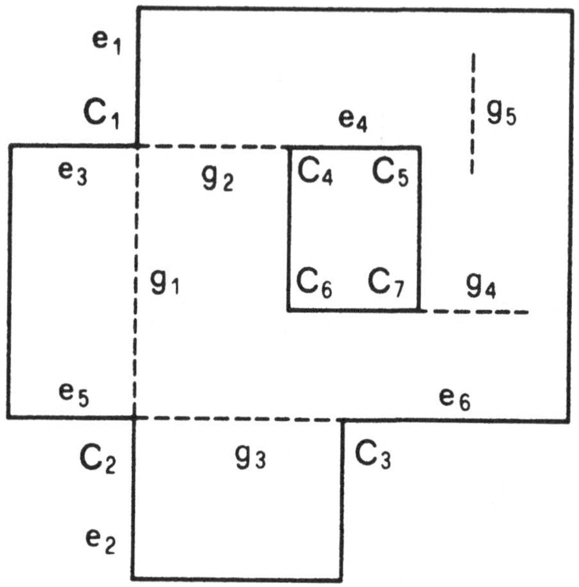

Fig. 3. A figure with a hole [5]. C_1 to C_7 are concave corners. g_1 to g_5 are lines. e_1-e_2, e_3-e_4, e_5-e_6 are couples of aligned edges. $\{g_1\}$ and $\{g_2, g_3\}$ are the complete sets of heavy lines.

horizontal (or vertical) line from each concave corner to the contour of F, as to form the contours of all the rectangles of MDD. Thus, the real difficulty arises whenever there are aligned edges in the figure.

We now define a *quasi minimal disjoint description* (QMDD) as the union of the MDD's for the disjoint figures obtained after inserting in F the heavy lines of a complete set S. Since such disjoint figures do not contain edges, their MDD's can be easily constructed as explained above. We also have [5]:

$$\#QMDD = \#(\text{concave corners}) - \#(\text{holes}) - \#S + 1.$$

These results can be verified in figure 4 (same example of figure 3), where we have formed the two QMDD's relative to the sets of heavy lines $S_1 = \{g_1\}$ and $\{g_2, g_3\}$.

It is quite intuitive, although non trivially provable, that any MDD is a QMDD. In fact, any MDD is a QMDD where the complete set of heavy lines has maximum cardinality. Denoting such a set by S_{max}, we have:

$$\#MDD = \#(\text{concave corners}) - \#(\text{holes}) - \#S_{max} + 1.$$

This is the case of figure 4.b.

Thus, the problem of finding MDD is conducted to the one of determining S_{max}. This problem can be solved in polynomial time by the use of a particular data structure, and through a known algorithm to find maximum matchings in bipartite graphs [14]. In fact we can form a bipartite graph whose two sets of nodes are respectively associated to the

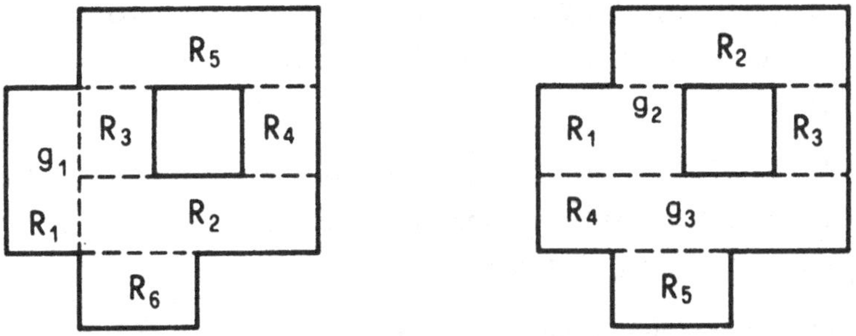

Fig. 4. Two QMDD's for the example of figure 3. We have in general: $\#QMDD = 7-1 - \#S_i + 1 = 7 - \#S_i$.
(a) Selecting $S_1 = \{g_1\}$ we have $\#S_1 = 1$, hence #QMDD=6.
(b) Selecting $S_2 = \{g_2, g_3\}$ we have $\#S_2 = 2$, hence #QMDD = 5. This QMDD is also an MDD.

sets of horizontal and vertical lines connecting adjacent edges, such that two nodes are connected by an arc if the two corresponding lines are incident. It is shown in [5] that S_{max} can be obtained from a maximum matching of the graph in polynomial time.

In conclusion, all the steps for the determination of an MDD of a given figure can be performed in polynomial time, by using a proper data structure.

REFERENCES

[1] E.LODI, F.LUCCIO, C.MUGNAI and L.PAGLI: A preliminary Investigation on two dimensional data organization. *Proc. Congresso A.I.C.A.*, 1976.

[2] S.P.GOSH: File organization: consecutive storage of relevant records on drum-type storage. *Information and Control* 25 (1974), 145-160.

[3] C.K.WONG: Data arrangement in magnetic bubble memories. *Proc. 15-th Allerton Conference*, 1977.

[4] E.LODI, F.LUCCIO, C.MUGNAI and L.PAGLI: On two dimensional data organization. *Fundamenta Informaticae* II (1978).

[5] W.LIPSKI, E.LODI, F.LUCCIO, C.MUGNAI and L.PAGLI: On two dimensional data organization II. *Fundamenta Informaticae* II (1978).

[6] W.J.MASEK: Some NP-complete set covering problems. *M.I.T. manuscript*, 1979.

[7] V.KLEE: Can the measure of $\cup_1^n [a_i, b_i]$ be computed in less than 0(n log n) steps? *American Math. Monthly* 84 (1977), 284-285.

[8] U.LAUTHER: 4-dimensional binary search trees as a means to speed up associative searches in design rule verification of integrated circuits. *J. of Design Automation and Fault-Tolerant Computing* 2 (1978), 241-247.

[9] J.L.BENTLEY and M.I.SHAMOS: Optimal algorithms for structuring geographic data. *Proc. Symp. on Topological Data Structures for Geographic Information Systems*, 1977.

[10] W.LIPSKI and F.P.PREPARATA: Finding the contour of a union of iso-oriented rectangles. *U. of Illinois manuscript*, 1979.

[11] W.LIPSKI: Finding a Manhattan path and related problems. *U. of Illinois manuscript*, 1979.

[12] F.LUCCIO and C.MUGNAI: Hamiltonian paths on a rectangular chessboard. *Proc. 16-th Allerto.. Conference*, 1978.

[13] D.E.KNUTH: Big omicron and big omega and big theta. *Sigact News* (1976), 18-24.

[14] J.E.HOPCROFT and R.M.KARP: An $n^{5/2}$ algorithm for maximum matchings in bipartite graphs. *SIAM J. on Computing* 2 (1973), 225-231.

COMPLEXITY OF OPTIMUM UNDIRECTED TREE PROBLEMS : A SURVEY OF RECENT RESULTS [(o)]

F. Maffioli
Istituto di Elettrotecnica ed Elettronica ,
Politecnico di Milano, 20133 Milano, Italy

Abstract

Networks design is often concerned with the problem of finding optimum weighted spanning trees. This work summarizes some recent results about the computational complexity of these problems with the aim of identifying the borderline between "easy" and "hard" problems. Several tree weight functions and side constraints are considered.

(o) Partially supported by an Italian Ministry of Education Research Contract.

" Varii gli effetti son, ma la pazzia
è tutt' una però, che li fa uscire.
Gli è come una gran selva, ove la via
conviene a forza, a chi vi va, fallire" [1]

1. Introduction

The theory of computational complexity[2,3,4] has provided fundamental insights into the inherent difficulty of many combinatorial optimization problems. Among these, problems concerned with tree structured networks are quite common. This work presents a survey of recent results towards a classification of optimum undirected tree problems from the point of view of their complexity (with respect to the worst case time norm), in order to identify the borderline between easy and NP-complete problems.

The general problem we are going to consider may be formulated as follows.

Instance: - an undirected graph $G = (V,A)$;
- arc weight functions $w_i : A \to R_i$, $i=1,2,\dots,k$;
- constants $W_i \in R_i$, $i=1,2,\dots,k$;
- a vertex $\rho \in V$;
- a subset of the vertex set $S \subseteq V$.

Question: is there a tree T of G spanning S such that $C_i(T) \leq W_i$, $i=1,2,\dots,k$?

In this formulation $C_i(.)$ are specified tree weight functions defined over the set $\mathcal{T}$ of all trees of arc weighted graphs and R_i is a weight range, such as $\mathbf{N}$, $\mathbf{Z}$, $\{-1\}$, $\{0,1\}$, etc.

Each problem is therefore identified by giving for $i=1,2,\dots,k$ a pair R_i, C_i. We use the following general notation:

$< R_1, C_1, \mid R_2, C_2, \mid \dots \mid R_k, C_k, >$.

Each problem may either be concerned with a spanning tree if $S \equiv V$, or with a Steiner-like tree if $S \subset V$.

The problems are always stated in "recognition" form (i.e. they require a yes - no answer), bearing in mind that the corresponding problems in "optimization" form are solvable in polynomial time if this is the case also for the problems in recognition form; similarly an optimization problem is "hard" if its recognition version is NP-complete[2,3,4].

2. General notations

We indicate by $V = \{v_1, v_2, \dots, v_n\}$ the set of vertices and by $A = \{a_1, a_2, \dots, a_m\}$ the set of arcs of an undirected graph $G = (V,A)$. The weight $w(a)$ of arc a and the weight $C(T)$ of a spanning tree T are always integer numbers, in order to avoid unnecessary technical complications arising from comparisons between real numbers.

The vertex $\rho \in V$ specified in the general problem formulation given in the introduction is also called root.

The path in a tree T of G from v_1 to v_2 is indicated by $\pi(v_1,v_2,T)$. The sum of the weights of the arcs in $\pi(v_1,v_2,T)$ is denoted by $p(v_1,v_2,T)$. The number of arcs in $\pi(v_1,v_2,T)$ is called the distance in T from v_1 to v_2.

The eccentricity $e(v)$ of a vertex v of a tree T is defined as the maximum distance in T from v to all the remaining vertices; the diameter of T is defined as the maximum eccentricity of its vertices. The valence of a vertex v of T is defined as the number of arcs incident to v. A leaf of T is a vertex of valence one. A branch at a vertex V of T is a maximal subtree containing v as a leaf. The balance of a vertex v of T is the maximum number of arcs contained in any branch at vertex v.

The number of paths which pass through an arc a in going from any vertex of $S_1 \subseteq V$ to any vertex of $S_2 \subseteq V$ in a tree T is indicated by $f_{S_1}^{S_2}(a,T)$. We also use the short-hand notations f(a,T) for $f_V^V(a,T)$ and $f_\rho(a,T)$ for $f_{\{\rho\}}^V(a,T)$: f(a,T) and $f_\rho(a,T)$ are called respectively the flow and the rooted flow in arc a. Moreover $\delta(a,T) = f(a,T)-w(a)$ and $\delta_\rho(a,T) = f_\rho(a,T)-w(a)$ are called respectively differential flow and rooted differential flow. Other notations are standard and may be found on several graph theory or combinatorial optimization books[15].

We conclude this Section by recalling the notion of reducibility among problems[2]. We say that problem P is reducible to problem P' (s.h.n. $P \propto P'$) if for any instance of P we may construct in polynomial time an instance of problem P' in such a way that the first instance is a yes-instance for P iff the second instance is a yes-instance for P'. The various symbols utilized to characterize the complexity of a problem are illustrated in Table 1.

Table 1

problem \ symbol	*	?	!
recognition form	easy	undecided	NP-complete
optimization form	easy	undecided	hard

3. Tree weight functions

The tree weight functions we are going to consider in this paper are listed in Table 2, where T denotes a spanning tree of an undirected graph G(V,A) with arc weight function $w : A \to R$.

Each function is defined in the first column and is identified by a name and an acronym in the second and third column respectively. Viewing these functions from the optimization (instead of recognition) point of view, acronyms beginning with M refer to bottleneck, or min-max problems, whereas acronyms beginning with Σ refer to min-sum problems.

Table 2

Tree weight function C(T)	Name	Acronym
$\max_{a \in T} w(a)$	MAX ARC	Ma
$\sum_{a \in T} w(a)$	SUM ARC	Σa
$\max_{a \in T} [w(a)\ f_\rho(a,T)]$	MAX ROOTED FLOW	Mrf
$\sum_{a \in T} w(a)\ f_\rho(a,T)$	SUM ROOTED FLOW	Σrf
$\max_{a \in T} [w(a)\ f(a,T)]$	MAX FLOW	Mf
$\sum_{a \in T} w(a)\ f(a,T)$	SUM FLOW	Σf
$\max_{a \in T} \delta(a,T)$	MAX DIFFERENTIAL FLOW	Md
$\max_{j \in V} p(\rho,j,T)$	MAX ROOTED PATH	Mrp
$\max_{i,j \in V} p(i,j,T)$	MAX PATH	Mp
$\max_{a \in T} \delta_\rho(a,T)$	MAX ROOTED DIFFERENTIAL FLOW	Mrd
$\sum_{j \in V} p(\rho,j,T)$	SUM ROOTED PATH	Σrp
$\sum_{i,j \in V} p(i,j,T)$	SUM PATH	Σp
$\max_{j \in V(T)} \sum_{a \text{ incident to } j \text{ in } T} w(a)$	MAXVALENCE	Mv
$\sum_{a \text{ a leaf of } T} w(a)$	SUM LEAF	$\Sigma \ell$

The lower case letters of each acronym identify the item to which the operator Max, or Σ applies. Note that any spanning tree problem having MAX ARC as tree weight function is obviously equivalent to the corresponding existance problem, i.e. the problem with $w = +\infty$. Tree weight functions involving flows or rooted flows are typical of problems concerning the minimization of communication costs between all the vertices of a tree, or between the root ρ and the remaining vertices. Tree weight functions Mp, Mrp are typical of problems where the maximum weighted distance (or travel time) between the vertices are of concern, as f.i. in emergency or highly reliable networks.

Weightings Md and Mrd are normally used in recognition problems with $W = 0$; these problems consist in finding a tree for which the (rooted) flow through each arc a does not exceed the capacity, w(a), of that arc. Finally, note that the functions Σp and Σrp are identical to Σf and Σrf respectively and have been reported in Table 2 for the sake of completeness.

4. Unconstrained spanning trees

The problems we are going to consider in this Section are characterized by $k = 1$ and $S \equiv V$.

We mention now some known results.

(i) The MAX ARC problem is obviously easy, and has been shown in[8] to be O(m). (Recall from Section 3 that this problem is equivalent to the problem of the existance in a graph of a spanning tree).

(ii) The SUM ARC problem with $w : A \to \mathbb{Z}$ is solvable in $O(n^2)$ time[18] or in O(m log log n) time[10] where m, n are respectively the number of arcs and vertices of the graph. The second method is better for sparse graphs.

(iii) Dijkstra's method[7] can be used to solve both the SUM ROOTED PATH problem and the MAX ROOTED PATH problem in $O(n^2)$ time, if the range of the arc weight functions is **N** .

(iv) The SUM PATH problem is nothing but what in[12] has been called Simple Network Design problem and proved to be NP-complete with $w : A \to \mathbf{N}$.

(v) The MAXVALENCE problem models very easily the hamiltonian path problem and is therefore NP-complete[2].

(vi) The SUMLEAF problem has been studied by Garey and Johnson[4] and is NP-complete even if the range of the arc weights is {1} and G is planar with valences not greater than 4.

The complexity results for unconstrained spanning tree problems are summarized in Table 3.

Table 3

Problem		
MAX ARC	*	*
SUM ARC	*	*
§MAX ROOTED FLOW	!	!
§SUM ROOTED FLOW (= Σrp)	*	!
MAX FLOW	!	!
§SUM FLOW (= Σp)	!	!
MAX DIFFERENTIAL FLOW	*	!
MAX ROOTED PATH	*	!
MAX PATH	!	!
§MAX ROOTED DIFFERENTIAL FLOW	!	!
§MAX VALENCE	!	!
§SUM LEAF	!	!

The results concerning MAX ARC, SUM ARC, SUM FLOW, MAX VALENCE and SUM LEAF problems are immediate consequences of points (i), (ii), (iv), (v) and (vi) respectively: in fact for any problem a "*" symbol in the third column implies a "*" symbol in the second, whereas "!" in the second column implies "!" in the third. Point (iii) accounts for the complexity of SUM ROOTED FLOW and MAX ROOTED PATH when the range is $\mathbf{N}$.

The remaining results of Table 3 are proved in[5].

A symbol § indicates that the arc weight range for which the corresponding problem has been shown to be NP-complete is more limited than what indicated, without proving all results of[5]. We report here as examples one of the NP-completeness proofs and the proof that < $\mathbf{N}$, MAX PATH > is easy.

MAX PATH is easy when the range of the arc weight functions is $\mathbf{N}$.

Proof. We need some preliminary definitions. The separation of a vertex v_i in an arc weighted tree T is $s(v_i,T) = \max_j p(v_i,v_j,T)$, where the maximum is taken over all vertices v_j of T.

Let $a = \{v_i,v_j\}$ be any arc in a tree T, $i < j$ and let x be any integer s.t. $0 \leq x \leq w(a)$.

Denote by $T_{a,x}$ the arc weighted tree obtained from T by

(i) adding a new vertex v_a and

(ii) substituting arc a with two arcs $a' = \{v_i,v_a\}$, $a'' = \{v_a,v_j\}$ with weights $w(a') = x$, $w(a'') = w(a)-x$.

We indicate with $\sigma(a,x,T)$ the separation of v_a in $T_{a,x}$, i.e.

$$\sigma(a,x,T) = s(v_a,T_{a,x}) .$$

Consider the following problem.

ABSOLUTE CENTRE LOCATION

Instance - arc weighted graph $G' = (V',A')$, $w' : A' \to \mathbf{N}$,
- $W \in \mathbf{N}$.

Question: do there exist:

- a spanning tree T' of G'
- an arc a' of T'
- a non negative integer $x' \leq w'(a')$

s.t. $\sigma(a',x',T') \leq W'$?

The problem stated above is the recognition version of an homonymous easy problem reviewed in[6]. Therefore it is sufficient to show that MAX PATH reduces to ABSOLUTE CENTRE LOCATION, whenever arc weight functions of MAX PATH range over the set of the even natural numbers. (It is straightforward to reduce MAX PATH with arc weight function ranging over $\mathbf{N}$ to the restricted version ranging only over the even naturals).

In fact let $G'=G, w'(a)=w(a)$ for all a A and $W' = \lfloor W/2 \rfloor$. Let T be a spanning tree of G s.t.

$$M = \max_{v_i, v_j \in V} p(v_i, v_j, T) = p(\mu, \nu, T) \leq W, \tag{1}$$

and let $P = \pi(\mu,\nu,T)$. Obviously, since M is even, there exists an arc $a = \{v_h, v_k\}$ of P and an integer x, $0 \leq x \leq w(a)$ s.t.

$$p(\mu, v_a, P_{a,x}) = p(v_a, \nu, P_{a,x}) = M/2 \tag{2}$$

Assume wlog that $\pi(\mu, v_a, P_{a,x})$ contains v_h and $\pi(v_a, \nu, P_{a,x})$ contains v_k. For each vertex v_ℓ of T other than μ,ν it must be that

$$p(v_\ell, v_a, T_{a,x}) \leq \frac{M}{2}. \tag{3}$$

In fact assume that for some ℓ, $p(v_\ell, v_a, T_{a,x}) > M/2$ and wlog that $\pi(v_\ell, v_a, T_{a,x})$ contains v_h. Then

$$p(v_\ell, \nu, T) = p(v_\ell, \nu, T_{a,x}) = p(v_\ell, v_a, T_{a,x}) + p(v_a, \nu, P_{a,x}) > \frac{M}{2} + \frac{M}{2} = M$$

in contradiction with (1).

From (2) and (3) it follows that

$$\sigma(a,x,T) = \max_{v_\ell \in V} p(v_a, v_\ell, T_{a,x}) = \frac{M}{2} \leq \left\lfloor \frac{W}{2} \right\rfloor = W'.$$

Conversely, suppose T' is a spanning tree of G' s.t. $\sigma(a',x',T') \leq W'$ where a' is an arc of T' and $x' \in \mathbf{N}$, $x' \leq w'(a')$. Then

$$M' = \max_{v_\ell \in V} p(v_{a'}, v_\ell, T_{a',x'}) \leq W' \tag{4}$$

Let v_i, v_j be any two vertices of T' and P_i, P_j be the set of arcs in $\pi(v_i, v_{a'}, T_{a',x'})$, $\pi(v_j, v_{a'}, T_{a',x'})$ respectively. Let $P_{i,j}$ be the set of arcs in $\pi(v_i, v_j, T_{a',x'})$. We have that

$$P_{i,j} = (P_i \cup P_j) - (P_i \cap P_j) ,$$

and since all arcs weights are non negative

$$p(v_i, v_j, T_{a',x'}) \leq p(v_i, v_{a'}, T_{a',x'}) + p(v_j, v_{a'}, T_{a',x'})$$

By (4) it follows that

$$p(v_i, v_j, T_{a',x'}) \leq 2\, W' \leq W \quad \square$$

MAX DIFFERENTIAL FLOW is NP-complete.

Proof. We start from the following NP-complete problem[2] :

EXACT 3-COVER (X3C)

Instance: A family $\mathcal{S} = \{\sigma_1, \ldots, \sigma_s\}$ of 3-element subsets of a set $\mathcal{T} = \{\tau_1, \ldots, \tau_{3t}\}$

Question: Does there exist a subfamily $\mathcal{S}' \subseteq \mathcal{S}$ of pairwise disjoint sets such that $\bigcup_{\sigma \in \mathcal{S}'} \sigma = \mathcal{T}$?

We exhibit a reduction of X3C to Md.

Let

$$V' = \mathcal{T} \cup \mathcal{S} \cup \{\phi_1, \phi_2\} \cup \{\rho_1, \rho_2, \ldots, \rho_r\}$$

where $r > 4t$,

$$A' = \{\{\sigma_i,\tau_j\} : \tau_j \in \sigma_i\} \cup \{\{\phi_1,\sigma_j\}, \forall j\} \cup$$

$$\{\{\phi_2,\sigma_j\}, \forall j\} \cup \{\{\phi_2,\rho_k\}, \forall k\}$$

and finally

$$w'(a') = \begin{cases} (4t+1)(s+r-t+1) & \text{if } a' = \{\phi_1,\phi_2\} \\ 4(n-4) & \text{if } a' = \{\phi_1,\sigma_j\}, \forall j \\ n-1 & \text{otherwise.} \end{cases}$$

An example of this reduction for $s = 4$, $t = 2$ is given in fig. 1. It is obvious that given a yes-instance of X3C, we can derive a yes-instance of Md (see wiggly lines of figure 1).

We prove now the converse. Let T' be a yes-instance of Md. Then for all σ_j, T' contains either $\{\phi_1,\sigma_j\}$ or $\{\phi_2,\sigma_j\}$, not both. In the second case σ_j is a leaf because of the constraint on the flow of arc $\{\phi_2,\sigma_j\}$. The τ_i's are for the same reason all leaves.

Let $Q = \{\sigma_j : \{\phi_1,\sigma_j\} \in T'\}$. We claim that $|Q| = t$ hence determining a yes-instance of X3C. Since the vertices τ_i can be attached only to vertices σ_j connected to ϕ_1 and not with those connected to ϕ_2, $|Q| \geq t$. Let $\Delta = |Q|-t$. We may write

$$f(\{\phi_1,\phi_2\},T) = (4t+1+\Delta)(s+r-t+1-\Delta)$$

$$= w'(\{\phi_1,\phi_2\})+\Delta(s+r-5t-\Delta).$$

But $\quad s+r-5t-\Delta > s-t-\Delta \geq 0$, for $r > 4t$ and $\Delta \leq s-t$.

Since $f(\{\phi_1,\phi_2\},T) \leq w'(\{\phi_1,\phi_2\})$, it follows that $\Delta = 0$, i.e. $|Q|=t$.

This proves the NP-completeness of the problem when **N** is the range of the weight function and hence also when the range is **Z** □

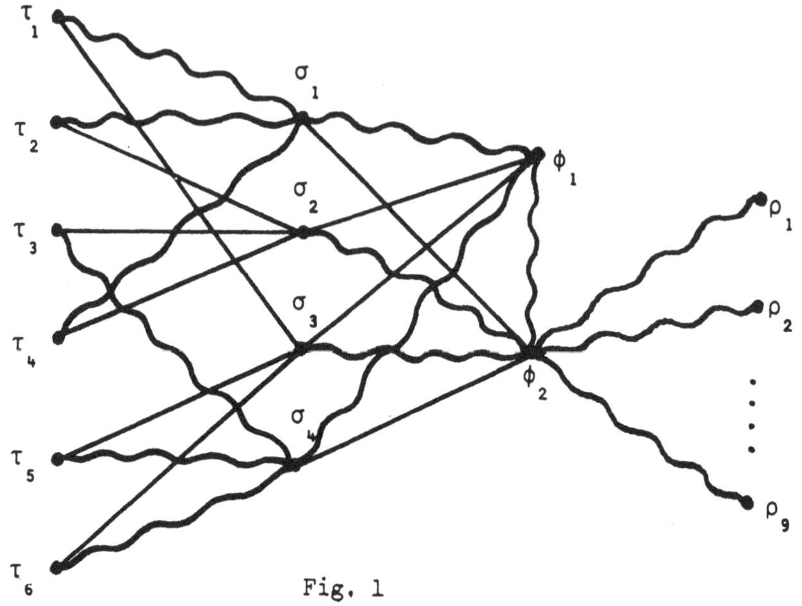

Fig. 1

5. Constrained problems

The constraints that we will take into account impose upper bounds on the following quantities defined on a spanning tree T:

(1) the eccentricity $e(\rho)$ of the root ρ,

(2) the diameter,

(3) the balance at ρ,

(4) the valences of the vertices,

(5) the number of leaves of T.

The first two constraints are typical of problems concerning the design of tree networks where the distance from a specified node or between every pair of nodes is modelling some kind of emergency or reliability requirement (see for instance[6]). Bounds on the balance of a root may be used to take into account capacity constraints on a central facility; applications of these kind of models are to be found in the design of centralized telecommunication networks[13]. It is certainly known to the reader how common is to require a bounded valence of the vertices of a network. Fig. 2 illustrates these constraints by an example.

All these constrained problems may be also considered in our general notation as <u>two criteria</u> problems. In fact let $R_2 = \{1\}$ and consider the following tree weight functions:

(1') MAX ROOTED PATH,

(2') MAX PATH,

(3') MAX ROOTED FLOW,

(4') MAX VALENCE,

(5') SUM LEAF.

The function (i') will model the corresponding constrained (i) problem.

$C_1(T)$ has been limited in every case at the set of tree weight functions which are easy without any further constraint: in fact, disregarding problems which are trivial to solve, a constrained problem is always at least as difficult as an unconstrained one.

As far as W_2 is concerned in some cases it has been decided to consider W_2 both as part of the instance of the problem, and as part of its name (W_2 = k fixed). The second class of problems is obviously easier than the first, and is indicated by a k introduced in the second field.

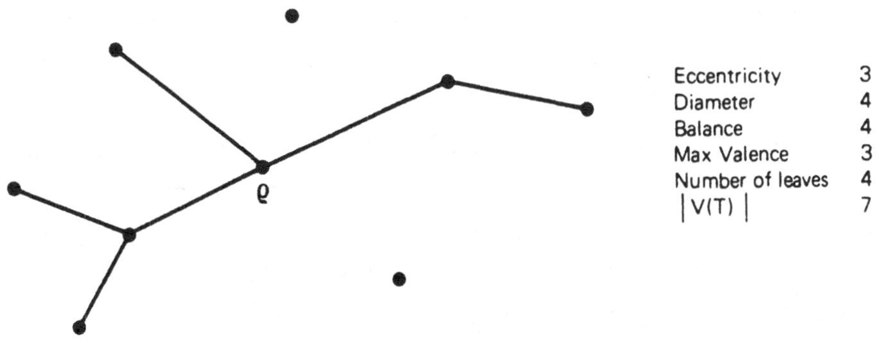

Fig. 2

5.1 Bounded eccentricity

In this Section we limit our attention to the following tree weight functions: Ma, Σa, Σrp, Mrp, Mrf, Mrd. In fact the concept of a root being implicit in the eccentricity constraint, it seemed more natural to consider rooted rather than unrooted tree weight functions.

Note that $\langle ., Ma|\{1\}, Mrp\rangle$ can be solved by weighting the arcs of the graph with a fixed positive integer and constructing the tree of shortest paths from the specified root ρ to the remaining vertices[7].

For all the remaining tree weight functions we have been able to show that the corresponding bounded eccentricity problems are in general NP-complete. More precisely we have the following results:

- $\langle \mathbb{N}, \Sigma a|\{1\}, Mrp, k\rangle$ is NP-complete for each $k \geq 2$,
- $\langle \mathbb{Z}, \Sigma rp|\{1\}, Mrp, k\rangle$ is NP-complete for each $k \geq 2$,
- $\langle \mathbb{N}, Mrp|\{1\}, Mrp, k\rangle$ is NP-complete for each $k \geq 3$,
- $\langle \mathbb{N}, Mrd|\{1\}, Mrp, k\rangle$ is NP-complete for each $k \geq 3$,
- $\langle \mathbb{N}, Mrf|\{1\}, Mrp, k\rangle$ is NP-complete for each $k \geq 3$.

The complexity of problem $\langle \mathbb{N}, \Sigma rp|\{1\}, Mrp, k\rangle$ remains to be decided for each $k \geq 2$. It is an easy matter to show that the same problems with smaller values of k are solvable in polynomial time.

Since this result shows that almost all the bounded eccentricity problems are NP-complete, in order to identify significant similar problems solvable in polynomial time we turned our attention to another restricted version of the eccentricity problems, namely the problems for which $W_2 \leq n-k$. These problems are similar to the preceeding problems with the only difference that they ask for the existance of a spanning tree of eccentricity less than or equal to n-k, with n being the number of vertices of the input graph and k being a given fixed integer. In this case we have the following result.

For each $k \geq 1$:

- $\langle \mathbf{Z}, \Sigma a | \{1\}, Mrp, n-k \rangle$ is easy ;
- $\langle \mathbf{N}, Mrp | \{1\}, Mrp, n-k \rangle$ and
  $\langle \mathbf{N}, \Sigma rp | \{1\}, Mrp, n-k \rangle$ are easy;
- $\langle \mathbf{Z}, Mrp | \{1\}, Mrp, n-k \rangle$ and
  $\langle \mathbf{Z}, \Sigma rp | \{1\}, Mrp, n-k \rangle$ are NP-complete;
- $\langle \mathbf{N}, Mrf | \{1\}, Mrp, n-k \rangle$ and
  $\langle \mathbf{N}, Mrd | \{1\}, Mrp, n-k \rangle$ are NP-complete.

5.2 Bounded diameter

For this constraint the results we have obtained up to now concern only Ma and Σa as far as $C_1(T)$ is concerned and are listed below.

- $\langle \mathbf{Z}, Ma | \{1\}, Mp \rangle$ is easy ;
- $\langle \mathbf{N}, \Sigma a | \{1\}, Mp, k \rangle$ is NP-complete for each $k \geq 4$;
- $\langle \mathbf{N}, \Sigma a | \{1\}, Mp, k \rangle$ is easy for $1 \leq k \leq 3$,
- $\langle \mathbf{N}, \Sigma a | \{1\}, Mp, n-k \rangle$ is easy for each $k \geq 1$.

5.3 Bounded balance

In this case we have the following result:

- $\langle \{1\}, Ma | \{1\}, Mrf \rangle$ is NP-complete for each $k \geq 3$,
- $\langle \{1\}, Ma | \{1\}, Mrf \rangle$ is easy for $k = 1,2$.

Notice that the first result is enough to conclude that bounded balance problems are NP-complete irrespective of the weighting function considered, except possibly when the value of k is fixed to 1 or 2.

5.4 Bounded valence

Since the problem of the existance of an hamiltonian path is well known to be NP-complete[2], it follows that $<\{1\},Ma|\{1\},Mv>$ is NP-complete.

A natural generalization of the bounded valence problems requires to find a spanning tree for which the valence of each vertex v_i is not greater than a given integer k_i, $i = 1,\dots,n$.

Whenever for some vertex v_i, $k_i \geq n-1$, the bound on the valence of v_i vanishes. In the particular case where no two adjacent vertices v_i, v_j exist s.t. both k_i and k_j are less than n-1, it is implicit in[14] that the corresponding bounded valence problem with the weight function $C_1 \equiv \Sigma a$ is easy.

5.5 Bounded number of leaves

The only result we may quote sofar can be found in[4] and in our notation states that $< \mathbf{N} , Ma|\{1\}, \Sigma\ell>$ is NP-complete.

The reduction is from the Dominating Set problem.

6. Steiner-like problem

6.1 Introduction

For these problems $S \subset V$. Two cases will be considered: unlabelled problems and labelled problems, referring to the fact that in the first case only the number of nodes to be spanned (i.e. $|S|$) is constrained ($S = \emptyset$), whereas in the second case the nodes to be spanned (terminal nodes) are precisely identified. Some general reductions are represented in fig. 3 and holds irrespectively from the tree weight function and the range considered for the problem.

It is worth observing that, if we assume $S = \emptyset$, the unlabelled problems with $S \geq k$ may be considered in our notation as double criteria problems of the kind

$$< R_1, C_1| \{-1\} , \Sigma a >$$

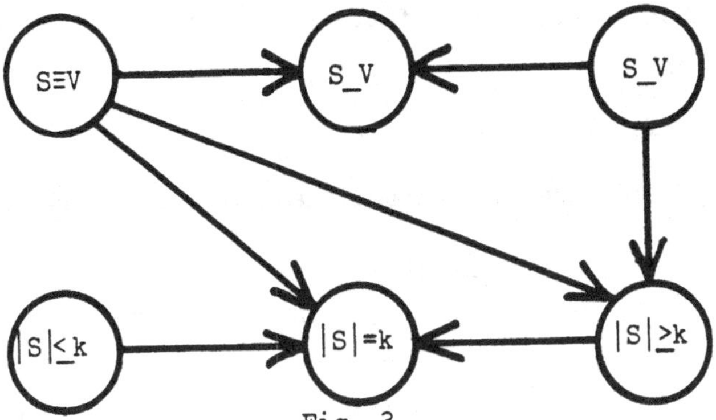

Fig. 3

The complexity results for these problems are reported in Table 4 and are proved as all other results of this Section in[9].

Table 4

Complexity of $\langle R_1, C_1 | \{-1\}, \Sigma a \rangle$ problems

C_1 \ R_1	N	Z
MAX ARC	*	*
SUM ARC	!	!
MAX ROOTED FLOW	!	!
SUM ROOTED FLOW	*	!
MAX FLOW	!	!
SUM FLOW	!	!
MAX DIFFERENTIAL FLOW	!	!
MAX ROOTED DIFF. FLOW	!	!
MAX ROOTED PATH	*	!
MAX PATH	*	!

As it appears also from fig. 3 problems for which $|S| = k$ are even more difficult than these, and in fact may be considered as triple criteria problems of the kind

$$< R_1, C_1 \mid \{-1\}, \Sigma a \mid \{1\}, \Sigma a > .$$

For these problems we have established the results of Table 5.

Table 5

Complexity of $< R_1, C_1 \mid \{-1\}, \Sigma a \mid \{1\}, \Sigma a >$ problems

C_1 \ R_1	**N**	**Z**
MAX ARC	*	*
SUM ARC	!	!
MAX ROOTED FLOW	!	!
SUM ROOTED FLOW	*	!
MAX FLOW	!	!
SUM FLOW	!	!
MAX DIFFERENTIAL FLOW	!	!
MAX ROOTED DIFF. FLOW	!	!
MAX ROOTED PATH	*	!
MAX PATH	*	!

The complexity of labelled problem is reported in Table 6.

Table 6

Complexity of labelled Steiner-like problems

Tree weight function \ Range	$\mathbb{N}$	$\mathbb{Z}$
MAX ARC	*	*
SUM ARC	!	!
MAX ROOTED FLOW	!	!
SUM ROOTED FLOW	*	!
MAX FLOW	!	!
SUM FLOW	!	!
MAX DIFFERENTIAL FLOW	!	!
MAX ROOTED DIFF. FLOW	!	!
MAX ROOTED PATH	*	!
MAX PATH	*	!

Many results follow from the general reduction of fig. 3 and from a well known NP-complete problem, i.e. the problem with tree weight function Σa (ND 12 in [4]). As an example of reduction we use this result to prove that $< \mathbb{N}\ , \Sigma a \mid \{-1\}, \Sigma a \mid \{1\}\ , \Sigma a\ >$ is NP-complete.

<u>Theorem</u> The Steiner-like problem $<\ , \Sigma a >$ (labelled) reduces to the unlabelled problem

$$< \mathbb{N}, \Sigma a \mid \{-1\}, \Sigma a \mid \{1\},\ \Sigma a\ >$$ which is therefore NP-complete.

<u>Proof</u> wlog assume

$$S = \{v_1, v_2, \ldots, v_{|S|}\}$$

and put

- $V' = V \cup \{u_1, u_2, \ldots, u_{|S|}\}$
- $\bar{A} = \{\{v_1, u_1\}, \ldots, \{v_{|S|}, u_{|S|}\}\}$
- $A' = A \cup \bar{A}$
- $G' = (V', A')$
- $W_1' = W + M(|S| + j - 1)$
- $k = 2|S| + j = W_2' = W_3'$

$\left.\right\}$ $j = 0, 1, \ldots, V - S$

$$- w_1'(a) = \begin{cases} w(a) + M & \text{if} \quad a \in A \\ 0 & \text{otherwise} \end{cases}$$

with M suitably large.

7. Conclusions

In this paper we have seen some unconstrained and constrained spanning tree problems from the viewpoint of their time complexity.

Two recent papers worth mentioning are concerned with two other kinds of constraints.

One[16] implies that the problem of the existance of a spanning tree obeying 2-parity constraints[15] is easy. The other[17] studies the complexity of spanning tree problems restricted to be isomorphic to some specific tree-structure, and succeeds in finding an easily computable function of the assigned structure, which captures the rate of growth of the complexity of the problem, provided some seemingly difficult to prove although probably true, conjectures are valid.

Some work in order to complete the overall picture about complexity of optimum tree problems is still to be done and will be the subject of part II of the paper of Camerini, Galbiati and this author[5].

Acknowledgment

The proof reported at the end of Section 4 is resulting from one of the many fruitful discussions with J.K. Lenstra. I am pleased to thank here him and A.H.G. Rinnooy Kan for these opportunities.

This whole work would not have been possible without the continuous collaboration of my friends P.M. Camerini and G. Galbiati.

References

1. Ariosto L., *Orlando Furioso* , XXIV, 2 (1532).

2. Karp, R.M., Reducibility among combinatorial problems, in: R.E. Miller and J.W. Thatcher (eds.), *Complexity of Computer Computations*, Pergamon Press, Oxford & N.Y. (1972) 85-103.

3. Aho, A.V., Hopcroft, J.E. and Ullman, J.D., *The Design and Analysis of Computer Algorithms,* Addison Wesley, Reading, Mass. (1974).

4. Garey, M.R. and Johnson, D.S., *Computers and Intractability,* W.H. Freeman & Co., S. Francisco (1979).

5. Camerini, P.M., Galbiati,G., and Maffioli, F., Complexity of Spanning tree problems: Part I, *EJOR* (to appear).

6. Christofides, N., *Graph Theory: an Algorithmic Approach,* Academic Press,London (1975), Ch. 5.

7. Dijkstra, E., A note on two problems in connection with graphs, *Numerical Mathematics,* 1 (1959) 269-271.

8. Camerini, P.M., The min-max spanning tree problem and some extensions, *Inf. Proc. Letters* 7 (1978) 10-14.

9. Camerini, P.M., Galbiati G., and Maffioli F., The complexity of Steiner-like problems , 17th Allerton Conf. on Communication, Control and Computing, October 1979.

10. Cheriton, D., and Tarjan, R.E., Finding minimum spanning trees, *SIAM J. on Computing,* 5 (1976) 724-742.

11. Cook, S.A., The complexity of theorem proving procedures, Proc. 3rd Annual ACM Symp. of Th. of Computing (1971) 151-158.

12. Johnson, D.S., Lenstra,J.K. , and Rinnooy Kan, A.H.G., The complexity of the network design problem, *Networks*, 8 (1978) 279-285.

13. Kershenbaum, A., Computing capacitated minimal spanning trees efficiently, *Networks*, 4 (1974) 299-310.

14. Lawler, E.L., Matroid intersection algorithms, *Mathematical Programming*, 9 (1979) 31-56.

15. Lawler, E.L., *Combinatorial Optimization: Networks and Matroids*, Holt, Rinehart and Winston, N.Y.,(1976).

16. Lovàsz, L., The matroid parity problem, Dept. of Combinatorics and Optimization, The University of Waterloo, 1979.

17. Papadimitriou, C.H., and Yannakakis, M., The complexity of restricted minimum spanning tree problems", Laboratory for Computer Sc., M.I.T., Cambridge, Mass., 1979.

18. Prim, R., Shortest connection network and some generalizations" *B.S.T.J.*, 36 (1957) 1389-1401.

AN INTRODUCTION TO POLYMATROIDAL NETWORK FLOWS

EUGENE L. LAWLER
Computer Science Division
University of California
Berkeley, CA 94720, USA

ABSTRACT

In the "classical" network flow model, flows are constrained by the capacities of individual arcs. In the "polymatroidal" network flow model, flows are constrained by the capacities of *sets* of arcs. Yet the essential features of the classical model are retained: the augmenting path theorem, the integral flow theorem, and the max-flow min-cut theorem all yield to straightforward generalization. In this paper we provide an introduction to the theory of polymatroidal network flows, with the objective of showing that this theory provides a satisfying generalization and unification of both classical network flow theory and much of the theory of matroid optimization.

This research was supported in part by NSF grant MCS78-20054, and in part by the Mathematisch Centrum, Amsterdam.

1. INTRODUCTION

In the "classical" network flow model, flows are constrained by the capacities of individual arcs. In the "polymatroidal" network flow model, flows are constrained by the capacities of *sets* of arcs. Yet the essential features of the classical model are retained: the augmenting path theorem, the integral flow theorem and the max-flow min-cut theorem all yield to straightforward generalization.

In this paper we provide an introduction to the theory of polymatroidal network flows. Our principal objective is to show that this theory provides a satisfying generalization and unification of both classical network flow theory and much of the theory of matroid optimization, including (poly)matroid intersection and matroid partitioning. We shall also indicate how the polymatroidal network flow model can be used to formulate and solve problems with no readily apparent polymatroidal structure.

The results presented here were obtained jointly with C.U. Martel [9], whose solution to a problem in multiprocessor scheduling suggested the formulation of the polymatroidal network flow model. It has come to our attention that the same model was formulated independently by Hassin [4]. A related model has also been investigated by Edmonds and Giles [3].

2. SOME POLYMATROIDAL PRELIMINARIES

We assume that the reader is familiar with the basic concepts of network flow theory and with at least some of the ideas of matroid optimization, as presented in [7]. In this section we present a few results concerning polymatroids which are needed in the remainder of the paper.

A *polymatroid* (E,ρ) is defined by a finite set of *elements* E and a *rank function* $\rho: 2^E \to \mathbb{R}^+$ satisfying the properties

$$\rho(\emptyset) = 0, \tag{2.1}$$

$$\rho(X) \leq \rho(Y) \quad (X \subseteq Y \subseteq E), \tag{2.2}$$

$$\rho(X \cup Y)+\rho(X \cap Y) \leq \rho(X)+\rho(Y) \quad (X \subseteq E,\ Y \subseteq E). \tag{2.3}$$

Inequalities (2.2) state that the rank function is monotone and inequalities (2.3) assert that it is submodular. If also ρ is integer-valued and $\rho(\{e\}) = 0$ or 1 for all $e \in E$, then the polymatroid is a *matroid*.

We shall be dealing with polymatroids whose elements are arcs of a network. We shall assign values of "flow" to these arcs, which is equivalent to specifying a function $f\colon E \to \mathbb{R}$. This function can be extended to subsets of E in a natural way, i.e.

$$f(\emptyset) = 0,$$
$$f(X) = \textstyle\sum_{x \in X} f(x) \quad (\emptyset \neq X \subseteq E). \tag{2.4}$$

Such an extended flow function f will be said to be *feasible* with respect to the rank function ρ if for all $X \subseteq E$,

$$f(X) \leq \rho(X). \tag{2.5}$$

A feasible function f *saturates* X if (2.5) holds with equality. An individual element e will be said to be *saturated* if there is some saturated set in which it is contained.

The following two lemmas apply with respect to any polymatroid (E,ρ) and any feasible function f.

LEMMA 2.1. *If* X *and* Y *are saturated sets, then so are* $X \cap Y$ *and* $X \cup Y$.

Proof. We have

$$\begin{aligned} f(X\cap Y) &\leq \rho(X\cap Y), && \text{by feasibility}\\ &\leq \rho(X)+\rho(Y)-\rho(X\cup Y), && \text{by submodularity}\\ &\leq f(X)+f(Y)-f(X\cup Y), && \text{by } f(X\cup Y) \leq \rho(X\cup Y) \text{ and saturation of } X,Y\\ &= f(X\cap Y), && \text{by (2.4)}. \end{aligned}$$

Hence $f(X\cap Y) = \rho(X\cap Y)$ and $X\cap Y$ is saturated. The proof for $X\cup Y$ is similar. □

LEMMA 2.2. *If* $e \in E$ *is saturated, then there is a unique minimal saturated set* $S(e)$ *containing* e. *Moreover, for each* $e' \in S(e)$, $e' \neq e$, *it is the case that* $f(e') > 0$.

Proof. Suppose $S(e)$ and $S'(e)$ are distinct minimal saturated sets containing e. By Lemma 2.1, $S(e) \cap S'(e)$ is also a saturated set containing e, and neither $S(e)$ nor $S'(e)$ can be minimal.

Now suppose $S(e)$ is the unique minimal saturated set containing e and there is an element $e' \neq e$ in $S(e)$ such that $f(e') = 0$.

$$
\begin{aligned}
f(S(e)-\{e'\}) &\leq \rho(S(e)-\{e'\}), && \text{by feasibility} \\
&\leq \rho(S(e)), && \text{by monotonicity} \\
&= f(S(e)), && \text{by assumption} \\
&= f(S(e)-\{e'\}), && \text{since } f(e') = 0.
\end{aligned}
$$

It follows that $S(e)-\{e'\}$ is also saturated and $S(e)$ cannot be the minimal saturated set containing e. □

3. POLYMATROIDAL FLOW NETWORKS

We shall consider only the simplest type of flow network, namely one in which there is a single *source* s and a single *sink* t. Our objective will be to find a maximum-value flow from s to t.

For each node j of the network there are specified two *capacity functions* α_j and β_j. The function α_j (β_j) satisfies properties (2.1)-(2.3) with respect to the set of arcs A_j (B_j) directed out from (into) node j. Thus (A_j,α_j) and (B_j,β_j) are polymatroids. (Comment: We permit there to be multiple arcs from one node to another. Hence A_j and B_j may be arbitrarily large finite sets.)

A *flow* in the network is an assignment of real numbers to the arcs of the network. We let a flow be represented by a function $f\colon 2^E \to \mathbb{R}$, obtained as in (2.4). A flow f is *feasible* if

$$f(A_j) = f(B_j), \qquad j \neq s,t, \tag{3.1}$$

$$f \text{ is feasible for } \alpha_j,\beta_j, \quad \text{for all nodes } j, \tag{3.2}$$

$$f(e) \geq 0, \qquad \text{for all arcs } e. \tag{3.3}$$

Equations (3.1) impose the customary flow conservation law at each node other than the source and sink. Property (3.2) indicates that capacity constraints are satisfied on sets of arcs, and (3.3) simply demands that the flow through each arc be nonnegative. Our objective is to find a feasible flow of maximum value, i.e. one which maximizes

$$v = f(A_s)-f(B_s) = f(B_t)-f(A_t). \tag{3.4}$$

If, for a given feasible flow f, the arc $e = (i,j)$ is saturated with respect to α_i, we shall say that the *tail* of e is saturated and denote the minimal saturated set containing e by $T(e)$, where $T(e) \subseteq A_i$. Similarly, if e is saturated with respect to β_j, we shall say that the *head* of e is saturated and denote the minimal saturated set containing e by $H(e)$, where $H(e) \subseteq B_j$.

In the case of an ordinary flow network in which there is a specified capacity c_{ij} for each arc $e = (i,j)$, we can define $\alpha_i(e) = \beta_j(e) = c_{ij}$, and then extend the functions α_i,β_i to sets as in (2.4). The resulting capacity functions are modular, i.e. satisfy (2.3) with equality. Note that in this special case the head of an arc e is saturated if and only if its tail is saturated, and $H(e) = T(e) = \{e\}$.

4. AUGMENTING PATHS

With respect to a given feasible flow f, an *augmenting path* is an undirected path of distinct arcs (but not necessarily distinct nodes) from s to t such that

(4.1) each backward arc e in the path is nonvoid, i.e. $f(e) > 0$, and

(4.2) if the head (tail) of a forward arc e in the path is saturated, then the following (preceding) arc in the path is a backward arc contained in $H(e)$ $(T(e))$.

In an ordinary flow network the minimal saturated set containing a saturated arc e is simply $\{e\}$, and since repetitions of arcs are not allowed, (4.2) does not permit any forward arc to be saturated. Thus, in this specialization our definition almost exactly coincides with the accepted notion of an augmenting path, the only (inconsequential) difference being that we permit repetitions of nodes.

We shall want to use augmenting paths in the customary way. That is, for some strictly positive δ, we want to increase the flow through each forward arc by δ and decrease the flow through each backward arc by δ, and thereby obtain an augmented flow which is feasible. It is not readily apparent that this can be done in our generalization.

LEMMA 4.1. *For any augmenting path there exists a strictly positive value of δ by which the flow can be augmented.*

Proof. There are two types of constraints on δ. First, the flow through each backward arc must remain nonnegative, and (4.2) assures us that there is a strictly positive value of δ for which this is possible. Second, for each node j and each $X \subseteq A_j$ (and similarly for each $X \subseteq B_j$) the resulting flow f' must be such that

$$f'(X) \leq \alpha_j(X).$$

Let $m(X)$ denote the number of forward arcs in X minus the number of backward arcs. Then we must have

$$f'(X) = f(X)+\delta m(X) \leq \alpha_j(X). \tag{4.3}$$

The only way in which (4.3) could fail to permit δ to be strictly positive would be for X to be saturated by f and for $m(X)$ to be strictly positive. But if X is saturated and contains forward arcs $e_1, e_2, \ldots, e_\ell$, then the tails of these forward arcs are saturated and $T(e_i) \subseteq X$, $i = 1,2,\ldots,\ell$. By (4.2), each e_i must be paired with a distinct backward arc $e_i' \in T(e_i)$. It follows that $m(X) \leq 0$, and the constraints (4.3) permit δ to be strictly positive. □

For many applications we need to be assured that there exists a maximal flow that is integer-valued. Hence we wish to obtain an integer version of Lemma 4.1.

An augmenting path can be *shortcut* if some portion of it can be removed to obtain a shorter augmenting path. For example, suppose an augmenting path contains two forward arcs e and e', both directed into the same node j, and that the heads of both of these arcs are unsaturated. If e occurs before e', then all the arcs following e up to and including e' can be removed from the augmenting path. A similar condition holds for two unsaturated forward arcs directed out from the same node.

The reader is invited to establish that an augmenting path which does not admit a shortcut has the following property: If the path passes through a given node j, the occurrences of j in the path are ordered as follows. First, there may be pairs of consecutive arcs of the form (e_h, e'_h), $h = 1,2,\ldots,k$, where each e_h is a forward arc directed into j whose head is saturated and $e'_h \in H(e_h)$. Second, there may be no more than one arc pair of the form (e,e'), where e is either a backward arc directed out from j or a forward arc directed into j whose head is unsaturated and e' is either a backward arc directed into j or a forward arc directed out from j whose tail is unsaturated. (If j = s(t), then there is only a single arc e' (e).) Third, there may be arc pairs of the form (e'_i, e_i), $i = 1,2,\ldots,\ell$, where each e_i is a forward arc directed out from j whose tail is saturated and $e'_i \in T(e_i)$. Moreover, the sets $H(e_h)$, {e,e'} and $T(e_i)$ are disjoint.

From these observations we can conclude that an augmenting path which does not admit a shortcut contains at most one forward arc in A_j which is unsaturated with respect to α_j and at most one forward arc in B_j which is unsaturated with respect to β_j. And, moreover, each of the sets $H(e_h)$ and $T(e_i)$ remains saturated after augmentation.

LEMMA 4.2. *Suppose all capacity functions and the existing feasible flow are integer-valued. Then for any augmenting path which admits no shortcut there exists a strictly positive integer value of δ by which the flow can be augmented.*

Proof. Let the maximum permissible value of δ be determined as in the proof of the previous lemma. If δ is determined by the amount of flow in a backward arc, then δ is an integer. So suppose a constraint of the form (4.3) is binding. If $m(X) = 1$, then δ is an integer. So suppose $m(X) > 1$. After augmentation of the existing flow f by δ, the resulting flow f' saturates X. As before, let $e_1, e_2, \ldots, e_\ell$ denote the forward arcs in X whose tails are saturated by f. Then the sets $T(e_i)$, $i = 1,2,\ldots,\ell$ remain saturated after augmentation, and the set

$$X' = X \cup T(e_1) \cup \ldots \cup T(e_\ell)$$

is also saturated by f'. Hence

$$f'(X') = f(X') + \delta m(X') = \alpha_j(X'). \tag{4.4}$$

But there is at most one forward arc in X' whose tail is unsaturated by f. Hence $m(X') \leq 1$ and (4.4) indicates that δ is integer. □

5. A LABELING PROCEDURE

Augmenting paths can be found by means of a labeling procedure which is much like that employed for ordinary flow networks. The principal difference is that labels are applied to arcs rather than to nodes. A labeling procedure which constructs augmenting paths without shortcuts is as follows:

Step 0. Initially all arcs are unlabeled and unscanned.

Step 1. To each nonvoid arc directed into s apply the label (-,*) and to each arc directed out from s whose tail is unsaturated apply the label (+,*).

Step 2. If there is an arc labeled "-" which is directed out from t or an arc labeled "+" which is directed into t whose head is unsaturated, stop. (An augmenting path has been found. The arcs in this path can be determined by backtracing, using the second component of each arc label.)

Step 3. If there are no arcs which are labeled and unscanned, stop. (There is no augmenting path.) Otherwise, find such an arc e and scan it as follows:
Suppose either e has a "+" label and is directed into node j or e has a "-" label and is directed out from node j. If e has a "+" label and its head is saturated, then apply the label (-,e) to all unlabeled arcs in H(e). If e has a "+" label and its head is unsaturated or e has a "-" label then apply the label (-,e) to all nonvoid arcs directed into j and apply the label (+,e) to all unlabeled arcs directed out from j whose tails are unsaturated. In addition, if e has a "-" label and its tail is saturated, apply the label (+,e) to all arcs e' such that $e \in T(e')$. Return to Step 2.

We have asserted in Step 3 that if the labeling procedure fails to find an augmenting path, then no augmenting path exists. This fact is by no means evident. The alert reader may even suspect that the labeling procedure may be defective, in that it permits a given arc to be given only one type of label ("+" or "-"), whereas both types might be applicable. We shall now prove that if the procedure fails to find an augmenting path then not only is there no augmenting path, but the flow is in fact maximal.

THEOREM 5.1 (Augmenting Path Theorem). *A flow is maximal if and only if it admits no augmenting path.*

Proof. If there is an augmenting path then Lemma 4.1 shows that the flow cannot be maximal. So suppose that the labeling procedure fails to find an augmenting path and let us show that this implies that the flow is maximal. The discussion which follows is with reference to the labels existing at the termination of the procedure.

Let us partition the nodes of the network into two sets, S and T. S is to contain node s, together with all nodes j such that either there is an arc directed from j with a "-" label or there is an arc directed into j with a "+" label whose head is unsaturated. All other nodes (including necessarily t) are in the set T.

We have thus defined a cut (S,T). Each "backward" arc (i,j), where $i \in T$, $j \in S$, must be void, else it would have received a "-" label and i would be in S. Let us partition the forward arcs (i,j), where $i \in S$, $j \in T$ into two sets U and L. Set U is to contain all unlabeled forward arcs and L is to contain all forward arcs which are labeled (either "+" or "-"). We thus have the situation indicated in Figure 1.

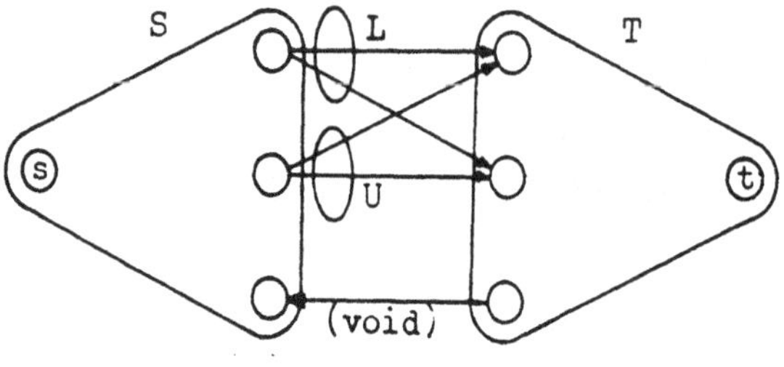

Figure 1. Cut (S,T,L,U).

Consider any node $i \in S$ and the set of arcs $U \cap A_i$. The tail of each arc $e \in U \cap A_i$ is saturated, else e could have a "+" label. Moreover, $T(e) \subseteq U \cap A_i$. For suppose there is some $e' \in T(e)$ such that $e' \notin U$. Such an arc e' cannot be unlabeled and directed to a node in S, else it could have received a "-" label. So e' must be labeled and directed to a node in T. If e' has a "-" label, then e could have received a "+" label from the scanning of e'. So e' must have a "+" label and this can be so only because there is some arc $e'' \in T(e')$ which has a "-" label. But if $e'' \in T(e)$ then e could have a "+" label. And if $e'' \notin T(e)$, we would have $e' \in T(e) \cap T(e') \neq T(e')$, a contradiction. Hence $e' \in U$ and $T(e) \subseteq U \cap A_i$. It follows that $U \cap A_i$ is the union of saturated sets and is itself a saturated set.

Now consider any node $j \in T$ and the set of arcs $L \cap B_j$. If an arc $e \in L \cap B_j$ has a "-" label, then there is a "+" labeled arc $e' \in L \cap B_j$ such that $e \in H(e')$. If an arc $e \in L \cap B_j$ has a "+" label, then its head is saturated (else $j \in S$) and $H(e) \subseteq L \cap B_j$, by the following reasoning. An arc $e' \in H(e)$ cannot be unlabeled because it could receive a "-" label from the scanning of e. If e' has a "-" label, then it must be directed from a node in S, by definition of S. If e' has a "+" label this label must have resulted from the scanning of an arc incident to a node in S. Hence $e' \in L$ and $H(e) \subseteq L \cap B_j$. It follows that $L \cap B_j$ is the union of saturated sets and is itself a saturated set.

We have shown that the net flow across the cut (S,T) is

$$\sum_{i \in S} f(U \cap A_i) + \sum_{j \in T} f(L \cap B_j) = \sum_{i \in S} \alpha_i(U \cap A_i) + \sum_{j \in T} \beta_j(L \cap B_j).$$

The flow is therefore maximal and there can be no augmenting path. □

From Theorem 5.1 and Lemma 4.2 we also obtain the following result in the case of integer capacities.

THEOREM 5.2 (Integral Flow Theorem). *If all capacity functions are integer-valued, then there is a maximal flow which is integral.*

6. MAX-FLOW MIN-CUT THEOREM

The proof of Theorem 5.1 clearly indicates the form of a max-flow min-cut theorem for polymatroidal network flows, which we now proceed to state.

An *arc-partitioned cut* (S,T,U,L) is defined by a partition of the nodes into two sets S and T, with $s \in S$, $t \in T$, and by a partition of the forward arcs across the cut into two sets U and L. The *capacity* of such an arc-partitioned cut is defined as

$$c(S,T,U,L) = \sum_{i \in S} \alpha_i(U \cap A_i) + \sum_{j \in T} \beta_j(L \cap B_j).$$

As in the case of ordinary flow networks, the value v of any feasible flow f is equal to the net flow across any cut, i.e.

$$v = f(U)+f(L)-f(B),$$

where B is the set of backward arcs, and clearly

$$v \le c(S,T,U,L). \tag{6.1}$$

THEOREM 6.1 (Max-Flow Min-Cut Theorem). *The maximum value of a flow is equal to the minimum capacity of an arc-partitioned cut.*

Proof. The proofs of Theorems 5.1 and 5.2, together with (6.1), are sufficient to establish the theorem for networks in which all capacity functions are integer (or rational) valued.

To complete the proof of the theorem, we must show that every network actually admits a maximal flow. (There is the possibility that a sequence of flow augmentations might fail to terminate with a well-defined maximal flow.) This question will be resolved in a later paper, in which questions of algorithmic efficiency will also be addressed. □

In the next several sections we shall indicate how the max-flow min-cut theorem specializes in various applications of the polymatroidal network flow model.

7. MATROID INTERSECTION

The (unweighted) matroid intersection problem is as follows. Given two matroids (E,ρ_1) and (E,ρ_2), find the largest possible set I which is independent in each of the matroids, i.e. such that $\rho_1(I) = \rho_2(I) = (I)$. This problem can be formulated and solved as a polymatroidal network flow problem, as shown in Figure 2. There are exactly two nodes, s and t, in the network and each arc from s to t corresponds to an element of E. The

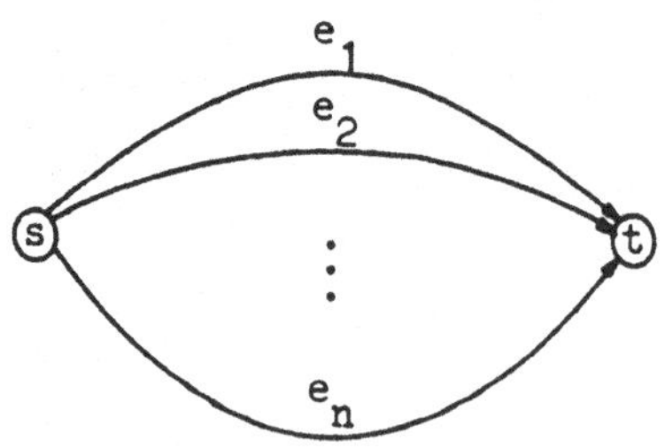

Figure 2. Flow Network for Matroid Intersection.

two capacity functions are determined by the two matroid rank functions: $\alpha_s = \rho_1$, $\beta_t = \rho_2$. Since these capacity functions are integer-valued, there exists a maximal flow which is integer. Any such integral maximal flow corresponds to a solution to the matroid intersection problem.

When the maximal flow algorithm suggested in Section 5 is applied to the network shown in Figure 2, it specializes precisely to the well-known matroid intersection algorithm [6,7]. An augmenting path without shortcuts corresponds to an "augmenting sequence". Minimal saturated sets T(e) and H(e) correspond to circuits $C_e^{(1)}$ and $C_e^{(2)}$ as defined in [6,7], and so forth.

It is also interesting to note that the max-flow min-cut theorem specializes exactly the well-known matroid intersection duality theorem. A partitioned cut (S,T,L,U) must have S = {s}, T = {t} and is obviously determined by a partition of the set E into subsets L and U. Thus we have the following

THEOREM 7.1 (Matroid Intersection Duality Theorem).

$$\max|I| = \min_{L\cup U=E} \{\rho_1(U)+\rho_2(L)\}.$$

8. MATROID PARTITIONING

Suppose we are given k matroids (E,ρ_i), i = 1,2,...,k, and we wish to determine whether or not there exists a partition of E into k sets I_i, i = 1,2,...,k, such that I_i is independent in (E,ρ_i). We can construct a flow network as shown in Figure 3. In this network each arc (s,e) has unit capacity, the flow into each node (E,ρ_i) is constrained by a capacity function $\beta_i = \rho_i$, and there are no other capacity constraints. If there exists an integral maximal flow of value |E| (which necessarily saturates each arc (s,e)), then there exists a partition of the desired type, otherwise not.

As it might be expected, when the maximal flow algorithm is applied to the network shown in Figure 3, it specializes to an algorithm very similar to that which has been proposed for solving the matroid partition problem [1,7].

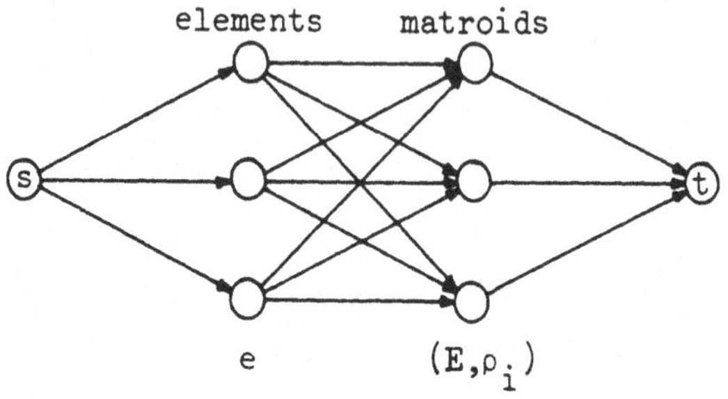

Figure 3. Flow Network for Matroid Partitioning.

A well-known set of necessary and sufficient conditions for the existence of a solution to the partitioning problem can be obtained quite easily from the max-flow min-cut theorem for polymatroidal network flows. There exists a solution to the matroid partitioning problem if and only if for the network of Figure 3 there does not exist an arc-partitioned cut with capacity strictly less than $|E|$. Any cut of finite capacity must be of of the form shown in Figure 4 where $S = A \cup \{s\}$, for some $A \subseteq E$. (A node (E,ρ_i) cannot be in S, else the cut would have unbounded capacity.) The capacity of such a cut is $\sum \rho_i(A) + |E-A|$, and if this is strictly less than $|E|$ we have $\sum \rho_i(A) < |A|$.

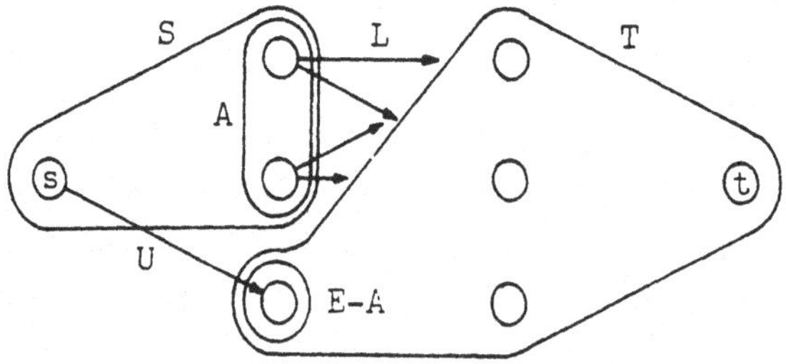

Figure 4. Cut in Proof of Theorem 8.1.

THEOREM 8.1 (Edmonds and Fulkerson [2]). *There exists a solution to the matroid partition problem if and only if for all* $A \subseteq E$,

$$|A| \leq \sum_i \rho_i(A).$$

9. A SCHEDULING PROBLEM

The polymatroidal network flow model can be applied to formulate and solve problems which have no readily apparent polymatroidal structure. The following problem in scheduling is such an example.

Suppose there are n *jobs*, $j = 1,2,\ldots,n$, each with a *release time* r_j, a *deadline* d_j, and a *processing requirement* p_j. It is desired to obtain a feasible preemptive schedule for these n jobs on m machines, where machine i has *speed* s_i, with $s_1 \geq s_2 \geq \ldots \geq s_m$. The usual conventions apply, i.e. a machine can work on only one job at a time and no job can be worked on by more than one machine at a time.

The set of 2n numbers $\{r_j\} \cup \{d_j\}$ defines at most 2n-1 distinct time intervals. We construct a flow network as shown in Figure 5, with a node for each job j and a node k for each of the time intervals. There is an arc (j,k) if and only if it is feasible to process job j in interval k. The arc flow f((j,k)) indicates the number of units of processing of job j to be done in interval k.

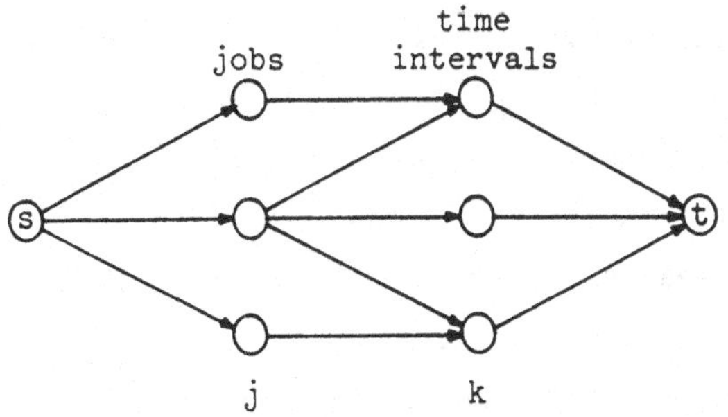

Figure 5. Flow Network for Scheduling Problem.

Suppose interval k has length t_k. It follows from a well-known result of scheduling theory that the arc flows into node k should be constrained by a capacity function of the form

$$\beta_k(X) = \begin{cases} (s_1+s_2+\ldots+s_\ell)t_k, & \text{if } |X| = \ell \le m-1, \\ (s_1+s_2+\ldots+s_m)t_k, & \text{if } |X| \ge m. \end{cases}$$

The function β_k is easily shown to satisfy properties (2.1)-(2.3).

Finally, a capacity of p_j is specified for each arc (s,j). There are no other capacity constraints in the network. We assert that there exists a feasible schedule if and only if there exists a flow whose value is equal to $\sum p_j$, i.e. a flow which saturates each arc (s,j).

If there does not exist a feasible schedule, the maximal flow algorithm identifies a subset of jobs for which it is easy to show that there is not enough available machine capacity. The details of the application of the max-flow min-cut theorem can be found in [9].

We comment that if the machines are *identical*, i.e. all have the same speed, the polymatroidal flow network for this problem specializes to an ordinary flow network [5]. If the machines are *unrelated*, i.e. have different speeds for different jobs, there appears to be no better alternative than solution by linear programming [8].

10. FURTHER EXTENSIONS

The polymatroidal network flow model can be extended and elaborated in many of the same ways as the classical model. Lower bounds on arc flow, in the form of supermodular set functions, can be applied. Convex cost functions can be specified for the flow through individual arcs, and a minimum-cost feasible flow can be computed. It appears that the polymatroidal model retains the desirable features of ordinary network flows under these extensions, and this will be one direction for future investigation.

REFERENCES

1. Edmonds, J., Minimum partition of a matroid into independent subsets, *J. Res. NBS*, 69B, 67-72, 1965.
2. Edmonds, J., Fulkerson, D.R., Transversals and matroid partition, *J. Res. NBS*, 69B, 147-153, 1965.
3. Edmonds, J., Giles, R., A min-max relation for submodular functions on graphs, *Annals Discrete Math.*, 1, 185-204,1977.
4. Hassin, R., On network flows, Ph.D. thesis, Yale University, May, 1978.
5. Horn, W., Some simple scheduling algorithms, *Naval Res. Logist. Quart.*, 21, 177-185, 1974.
6. Lawler, E.L., Matroid intersection algorithms, *Math. Programming*, 9, 31-56, 1975.
7. Lawler, E.L., *Combinatorial optimization: networks and matroids*, Holt, Rinehart and Winston, 1976.
8. Lawler, E.L., Labetoulle, J., On preemptive scheduling of unrelated parallel processors by linear programming, *J. ACM*, 25, 612-619, 1978.
9. Martel, C.U., Generalized network flows with an application to multiprocessor scheduling, Ph.D. thesis, University of California at Berkeley, May, 1980.

Approximation Algorithms for Bin Packing Problems: A Survey

M. R. Garey and D. S. Johnson
Bell Laboratories
Murray Hill, New Jersey 07974

Abstract. *Bin packing problems, in which one is asked to pack items of various sizes into bins so as to optimize some given objective function, arise in a wide variety of contexts and have been studied extensively during the past ten years, primarily with the goal of finding fast "approximation algorithms" that construct near-optimal packings. Beginning with the classical one-dimensional bin packing problem first studied in the early 1970's, we survey the approximation results that have been obtained for this problem and its many variants and generalizations, including recent (unpublished) work that reflects the currently most active areas of bin packing research. Our emphasis is on the worst-case performance guarantees that have been proved, but we also discuss work that has been done on expected performance and behavior "in practice," as well as mentioning some of the many applications of these problems.*

1. Introduction

Bin packing is a problem that has shown up under many guises, but to the authors of this survey, it is mainly known as "The Problem That Wouldn't Go Away." Every year or so we finish a paper on the subject, heave a sigh of relief, and remark that at long last we can wash our hands of bin packing. But, inevitably, some member of an ever-growing band of researchers on the topic comes up with yet another variant on the model, a new question to ask, or even (heaven forbid) a new application, and we find ourselves charging back into the fray for "just one last time." We have been back and forth over the territory so often now that it seems appropriate for us to prepare a new map. (Some earlier maps can be found in [25,33,34,35]). We hope this survey will be of use in pulling together many of the disparate strands that have (often in disguised form) made up the fabric of bin packing research to date.

The classical one-dimensional bin packing problem can be stated as follows: Suppose we are given a positive integer bin capacity C and a set or *list* of items $L=(p_1,p_2,\cdots,p_n)$, each item p_i having an integer size $s(p_i)$ satisfying $C \geq s(p_i) \geq 0$. What is the smallest integer m such that there is a partition $L=B_1 \bigcup B_2 \bigcup \cdots \bigcup B_m$ satisfying $\sum_{p_i \in B_j} s(p_i) \leq C$, $1 \leq j \leq m$? We usually think of each set B_i as being the contents of a *bin* of capacity B, and view ourselves as attempting to minimize the number of bins needed for a packing of L.

This abstract problem can be used to model a variety of problems in the real world, from packing trucks having a given weight limit to assigning commercials to station breaks on television [7]. One commonly cited example is the plumber's pipe-cutting problem: A plumber needs a collection of pipes of lengths $s(p_1)$, $s(p_2)$, ... , $s(p_n)$, which she can obtain by cutting up standard 8-foot lengths ($C=8$) that she purchases at her local plumbing supply store. She wishes to buy as few of these 8-foot lengths as possible.

Unfortunately for the Teamsters and the television network executives (but fortunately for the plumbing supply stores), finding optimal solutions for the bin packing problem is quite difficult. In fact, it, or more precisely the decision problem "Given C, L, and an integer bound K, can L be packed into K or fewer bins of capacity C?" is NP-complete. By the theory elaborated in [26,40], this means that it is unlikely that any efficient, (i.e., polynomial time) optimization algorithm can be found for the problem. Thus researchers have turned to the study of approximation algorithms for this problem, that is, algorithms which, although not guaranteed to find an optimal solution for every instance, attempt to find *near*-optimal solutions.

The bin packing problem holds a special place in the history of approximation algorithms. It was in this context that some of the first work was done in proving that fast approximation algorithms could actually *guarantee* near-optimal solutions. Earlier work by R. L. Graham [31,32] on certain multiprocessor scheduling problems was seen in retrospect to have done the same thing, but it was the bin packing results that served to popularize the approach. Consider the following simple procedure, called NEXT FIT: Process the items in L in turn, starting with p_1, which is placed in bin B_1. Suppose that p_i is now to be packed, and let B_j be the highest indexed non-empty bin. If p_i will fit in B_j (the sum of the sizes of items in B_j does not exceed $C-s(p_i)$), then put p_i in bin B_j. Otherwise, start a new bin (bin B_{j+1}) by putting p_i into it.

This is clearly a fast algorithm (linear time). Moreover, it is not difficult to show that, if NF(L) is the number of bins used in the NEXT FIT packing of list L and OPT(L) is the number of bins required in an optimal packing, then for all lists L,

$$\mathrm{NF}(L) \leq 2 \cdot \mathrm{OPT}(L)$$

This is the best bound of this sort that we can prove for NEXT FIT, since the examples shown in Fig. 1 indicate that there are lists L with $\mathrm{NF}(L) \geq 2 \cdot \mathrm{OPT}(L) - 1$.

To improve on this bound we need a new algorithm. One defect of NEXT FIT seems to be that it only tries to put p_i in one bin before it resorts to starting a new bin. This suggests that the following FIRST FIT algorithm might be an improvement: When packing p_i, put it in the lowest indexed bin into which it will fit (starting a new bin only if p_i will not fit into any non-empty bin). It can be shown (though the proof [24,39] is more difficult) that for all lists L,

$$L = (\frac{1}{2}, \frac{1}{2N}, \frac{1}{2}, \frac{1}{2N}, \ldots, \frac{1}{2}, \frac{1}{2N})$$

N bins   2N bins

OPT(L) = N + 1   NF(L) = 2N

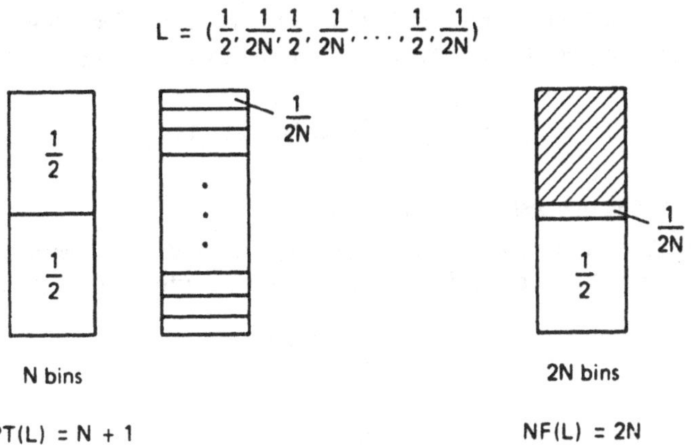

Figure 1. *Examples of lists L with NF(L) = 2·OPT(L) − 1.*

$$FF(L) \leq \frac{17}{10} \cdot OPT(L) + 1$$

and, again, this is the best ratio possible, since there are lists L with arbitrarily large values of OPT(L) such that FF(L) ≥ (17/10)·OPT(L) − 1 [39]. These lists are too complicated to illustrate here, but Fig. 2 shows examples which approach a ratio of 5/3 = 1.6666...

$$L = (\frac{1}{6} - 2t, \ldots, \frac{1}{6} - 2t, \frac{1}{3} + t, \frac{1}{2} + t, \ldots, \frac{1}{2} + t)$$

6N bins   N bins   3N bins   6N bins

OPT(L) = 6N   FF(L) = 10N

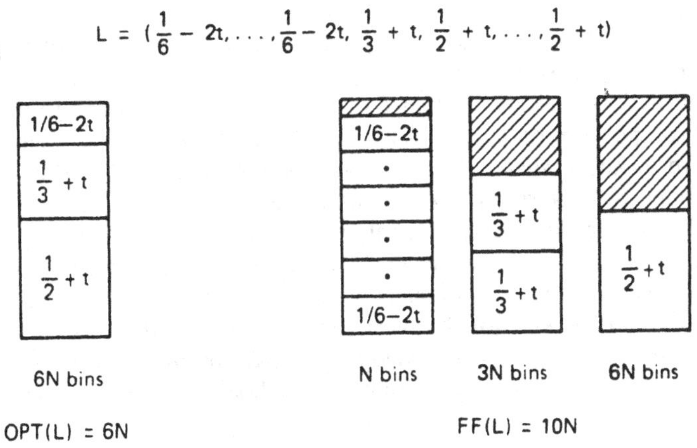

Figure 2. *Examples of lists L with FF(L) = $\frac{5}{3}$·OPT(L).*

From these examples a further improvement suggests itself. FIRST FIT seems to get into trouble when the large items occur at the end of the list. The algorithm FIRST FIT DECREASING seeks to avoid this by first ordering the items so that $s(p_1) \geq s(p_2) \geq \cdots \geq s(p_n)$, and then applying FIRST FIT to the reordered list. For this algorithm it can be shown (with considerable difficulty [37,39]) that for all lists L,

$$\mathrm{FFD}(L) \leq \frac{11}{9} \cdot \mathrm{OPT}(L) + 4$$

and, once more, this is the best ratio possible.

Let us formalize the type of worst case analysis we have been discussing. If A is an algorithm and A(L) is the number of bins used by that algorithm for list L, define

$$R_A(L) \equiv \frac{\mathrm{A}(L)}{\mathrm{OPT}(L)}$$

The *absolute performance ratio* R_A for algorithm A is given by

$$R_A \equiv \inf\{r \geq 1: R_A(L) \leq r \text{ for all lists } L\}$$

The *asymptotic performance ratio* R_A^∞ for A is given by

$$R_A^\infty \equiv \inf\{r \geq 1: \text{for some } N>0,\ R_A(L) \leq r \text{ for all } L \text{ with } OPT(L) \geq N\}$$

The above results can now be summarized by saying that $R_{NF}^\infty = 2$, $R_{FF}^\infty = 17/10$, and $R_{FFD}^\infty = 11/9$. Notice that R_A need not equal R_A^∞. Although $R_{FFD}^\infty = 11/9$, it is easy to give lists L for which OPT(L)=2 and FFD(L)=3, so that $R_{FFD} \geq 3/2$. The asymptotic ratios seem to be a more reasonable measure of performance for the basic bin packing problem, but absolute ratios do come up in some of the work on related problems that we shall be discussing later.

Another measure of performance that is relevant to the analysis of approximation algorithms is *expected*, or *average case*, behavior. In practice one would certainly like to have some idea of the "typical" performance of an algorithm in addition to the knowledge of how poorly it can perform on some possibly quite atypical worst case examples. Unfortunately, it is hard to know what a "typical" bin packing instance is – this will no doubt vary from application to application. As a consequence of this fact (or, more likely, as a consequence of the fact that probabilistic analysis can be incredibly difficult for all but the simplest algorithms), the average case analysis literature for bin packing is much sparser that that for worst case analysis. Although we shall mention average case results when they are known, our survey will, for the most part, be dominated by the worst case point of view. In the absence of precise knowledge of what constitutes a "typical" instance, there is considerable value in knowing guarantees that hold for all instances, including the typical ones.

Our survey will be structured as follows: Section 2 concentrates on the classical bin packing problem discussed above, elaborating on the algorithms already presented and their performance, covering a variety of alternatives that have been proposed, and pointing out what is known about the theoretical limitations of the various approaches.

Section 3 discusses variants on the classical problem in which additional restrictions are placed on the structures of allowable packings, restrictions that come up in such applications as multiprocessor scheduling, assembly line balancing, and dynamic storage allocation.

Section 4 discusses variants on the classical problem in which the objective function is altered, for instance by fixing the number of bins and attempting to maximize the number of items packed or to minimize the bin capacity needed so that all items can be packed.

Section 5 considers a generalization of one-dimensional bin packing to the multi-dimensional problem of packing vectors, so that the vector sum of items in an individual bin has no component exceeding C.

Section 6 considers a different way of generalizing to higher dimensions. The problem here is that of packing rectangles into rectangular bins or strips. This is quite an active area of current research, due to applications to VLSI (very large scale integrated circuit) design.

Indeed, research continues into many of the problems we shall be discussing, and there is no guarantee that some of the results we cite may not be out of date by the time you read this. In Section 7 we conclude by pointing out some of these "hot" areas and also suggesting other areas where interesting questions lie unanswered.

2. Classical Bin Packing

In this section we elaborate on the results known for the classical bin packing problem. Table 1 highlights the results that have been obtained. The quantity $R_A^\infty(t)$, $0<t\leq 1$ is the asymptotic worst case ratio for algorithm A on lists all of whose items have size bounded by $t \cdot C$. This measure is of interest in applications where the largest item expected is significantly smaller than the bin capacity.

The algorithms REVISED FIRST FIT and MODIFIED FIRST FIT DECREASING are recent developments which we shall be discussing in detail shortly. Most of the other results in the table were already known by 1973 [37,38]. The algorithm BEST FIT (BF) is like FIRST FIT, except that p_i is placed in the bin into which it will fit with the smallest gap left over (with ties broken in favor of the lowest indexed bin) [39]. WORST FIT (WF) [37,38] places p_i in the non-empty bin with the biggest gap (ties broken in the same way), starting a new bin if this biggest gap is not big enough. ALMOST WORST FIT (AWF) [37,38] tries the second largest gap first, and then proceeds as does WORST FIT – surprisingly, this makes a difference. The analysis of NEXT FIT DECREASING has been done recently by Baker and Coffman [4]. GROUP FIT GROUPED (GFG) [37,38] uses "implicit rounding" to discretize the ranges of item sizes and bin levels, thus avoiding the sorting implicit in the FFD algorithm which it mimics and attaining a linear running time, while paying only a partial penalty in worst case behavior. FIRST FIT GROUPED (FFG) [37,38] is a hybrid algorithm, included mainly because it yields a different value of R_A^∞. The algorithm ITERATED LOWEST FIT DECREASING, which attains the same value, but more slowly, is due to Krause, Shen, and Schwetman [44], and will be discussed in more detail in the next section.

A variety of other results were also obtained during the early 1970's. The precise

Algorithm	Timing	R_A^∞	$R_A^\infty(1/2)$	$R_A^\infty(1/3)$	$R_A^\infty(1/4)$
WORST FIT	$\theta(n\log n)$	2.0	2.0	1.5	1.333...
NEXT FIT	$\theta(n)$	2.0	2.0	1.5	1.333...
FIRST FIT	$\theta(n\log n)$	1.7	1.5	1.333...	1.25
BEST FIT	$\theta(n\log n)$	1.7	1.5	1.333...	1.25
ALMOST WORST FIT	$\theta(n\log n)$	1.7	1.5	1.333...	1.25
NF DECREASING	$\theta(n\log n)$	1.691...	1.424...	1.302...	1.234...
REVISED FF	$\theta(n\log n)$	1.666...	NA	NA	NA
GROUP FIT GROUPED	$\theta(n)$	1.5	1.333...	1.25	1.20
FF GROUPED	$\theta(n\log n)$	1.333...	1.333...	1.25	1.20
ITERATED LFD	$\theta(n\log^2 n)$	1.333...	NA	NA	NA
FF DECREASING	$\theta(n\log n)$	1.222...	1.183...	1.183...	1.15
BF DECREASING	$\theta(n\log n)$	1.222...	1.183...	1.183...	1.15
MODIFIED FFD	$\theta(n\log n)$	[1.183...,1.20]	1.183...	1.183...	1.15

Table 1. Asymptotic worst case bounds for bin packing algorithms.

values of $R_A^\infty(t)$ as a function of t were obtained for many of the algorithms [37,38,39]. Except for the algorithms WORST FIT and NEXT FIT, which yield the continuous function $R_A^\infty(t) = 1 + t/(1-t)$, these tend to be step functions determined by $\lfloor 1/t \rfloor$. In [37] the asymptotic worst case behavior of FIRST FIT was completely determined for the case when all item sizes lie in a specified interval, as a function of the interval. The algorithms NEXT-k FIT, $k \geq 1$, which resemble NEXT FIT except that p_i is placed in a new bin only if it will not fit in any of the last k non-empty bins (NEXT-1 FIT is the same as NEXT FIT), were studied in [37,38]. This paper also analyzed what might be considered "nondeterministic" bin packing algorithms: ANY FIT (AF), which can place p_i anywhere, except that it can never put it in a new bin unless it won't fit in any of the already non-empty bins, and ALMOST ANY FIT (AAF), which in addition can never put p_i in a bin whose gap is larger than that of all other bins unless that is the only place it fits. The results for ALMOST ANY FIT are the same as those for FIRST, BEST, and ALMOST WORST FIT (all of which obey the AAF ground rules), while the results for ANY FIT are the same as those for WORST FIT, which essentially makes the worst choices allowed under the AF ground rules. ANY FIT DECREASING and all DECREASING algorithms obeying the ANY FIT ground rules seem to obey the same bounds as FFD, although the best that has been proved for any such algorithms (other than FFD and BFD) is that $R_A^\infty \leq 5/4 = 1.25$ [37,38].

We should also point out the important work of Gilmore and Gomory in the early 1960's [28,29] for the case when the *number*, rather than range, of possible item sizes is limited. They were able to show that in this case the linear programming relaxation of the problem, although still quite large (it has a variable for each combination of possible item sizes that could fit together in a bin), can be solved using special techniques. An actual packing is then constructed by "rounding up" the solution values. In terms of worst case analysis, this algorithm will have $R_A^\infty = 1$ for any fixed number of item sizes, since it can yield at most one excess bin for each possible bin type (a much larger, but still fixed number, independent of the number of items). We note in passing that when the number of item sizes is fixed, we actually can find *optimal* solutions in polynomial time, although the degree of the polynomial can be astronomical (exhaustive search will do the trick). Gilmore and Gomory's contribution is in obtaining *almost* optimal solutions with much less work. Despite its inherent exponentiality when the number of item sizes is not limited. their approach has proven to be a very practical one in many applications where such limits are imposed, and can be adapted to many variants on the classical problem.

Renewed interest in the unrestricted classical problem has been inspired by a recent paper by A. C. Yao [53]. In the first part of this paper, the question of "on-line" algorithms is considered. An *on-line* algorithm is one which, like NEXT FIT or FIRST FIT, assigns items to bins in exactly the order they are given in the original list, without using any knowledge about subsequent items in the list. FIRST FIT DECREASING, for example, is not an on-line algorithm, since it first re-orders the list. On-line algorithms may be the only ones that can be used in certain situations, where the items to be packed are arriving in a sequence according to some physical process and must be assigned to a bin as soon as they arrive. Although it is known that "off-line" algorithms such as FFD can do much better than FIRST FIT, the question arises: "What is the best an on-line algorithm can do?" Until Yao's work, FIRST FIT was the best on-line algorithm known, with $R_{FF}^\infty = 1.7$. On the basis of a clever analysis of the worst case examples for FIRST FIT, Yao was able to devise a new algorithm, REVISED FIRST FIT (RFF), with $R_{RFF}^\infty = 5/3 = 1.6666$. Even more significantly, he was able to show that for *any* on-line algorithm A, we must have $R_A^\infty \geq 1.5$. (Subsequently, D. Brown [8] and F. M. Liang [45] have independently extended Yao's lower bound results, improving the lower bound to 1.536. In addition, Brown has recently designed a doubly revised FIRST FIT algorithm, whose asymptotic worst case ratio is better than 1.64 [9]).

In the second half of [53], Yao moves on to the challenge of improving on FIRST FIT DECREASING, which had long reigned as the bin packing champ, with 11/9 being the best value of R_A^∞ known for any polynomial time algorithm A. A number of challengers had been proposed previously [33,37], including one non-polynomial algorithm that optimally packs the first bin and then proceeds down the line, optimally packing each bin in turn, but all of these can be shown to be significantly worse than FFD. Yao once again devised an improved algorithm, this time based on a clever analysis of the proof of the FFD bound. The new algorithm, REVISED FIRST FIT DECREASING, runs in polynomial time – $O(n^{10}\log n)$, and has

$$R_{RFFD}^\infty \leq \frac{11}{9} - \frac{1}{10{,}000{,}000}$$

Of course, even before this result, there was no strong evidence that 11/9 was an absolute bound on what could be obtained by a polynomial time algorithm. Indeed, it would be consistent with the NP-completeness results proved so far for a polynomial time algorithm A to guarantee $A(L) \le \mathrm{OPT}(L)+1$ for all lists L. However, it took Yao's result to inspire us to look seriously at the question of beating FFD ourselves, and recently we have come up with an algorithm which is a much more significant improvement [27]. We call our algorithm MODIFIED FIRST FIT DECREASING (to distinguish it from Yao's "REVISED" algorithm), and so far have been able to verify that

$$\frac{71}{60} = 1.18333... \le R^{\infty}_{MFFD} \le 1.20 = \frac{6}{5}$$

The algorithm is based on a careful analysis of the 11/9 examples for FFD and what causes FFD to mispack them. It proceeds as follows: Partition the input list L into three sublists -

$$L_A = \{p_i: s(p_i) \in (\frac{C}{3}, C]\}$$

$$L_D = \{p_i: s(p_i) \in (\frac{C}{6}, \frac{C}{3}]\}$$

$$L_X = \{p_i: s(p_i) \in (0, \frac{C}{6}]\}$$

Our first step is to pack the sublist L_A using FFD. In the resulting packing, we call a bin containing only a single item from L_A an "A-bin." We then attempt to pack as much of L_D into A-bins as possible using the following loop:

1. Let bin B_j be the A-bin with the currently largest gap. If the two smallest unpacked items in L_D will not fit together in B_j, exit from the loop.
2. Let p_i be the smallest unpacked item in L_D, and place p_i in B_j.
3. Let p_k be the *largest* unpacked item in L_D that will now fit in B_j, and place p_k in B_j. Go to 1.

The assignment of items to bins is then completed by combining the unpacked portion of L_D with L_X and adding all these remaining items to the packing using FFD.

In addition to being an improvement over Yao's algorithm in asymptotic worst case ratio, MFFD also has a claim to being practical, since it can be implemented to run in time $O(n\log n)$. However, if all items are bounded by $C/2$, MFFD will construct exactly the same packings as FFD. Hence $R^{\infty}_{MFFD}(1/2) = R^{\infty}_{FFD}(1/2) = 71/60$ and so here Yao is still the reigning champion, since in [53] he gives an $O(n\log n)$ algorithm with $R^{\infty}_{A}(1/2) = (71/60) - (1/100{,}000)$.

It should be noted that all this work on surpassing FFD relies on the proof methods used in the original "11/9 Theorem," which ran on for 100 pages or so in [37]. Thus it may be a good omen for future results that at long last a "short" proof of that old result has been found, with B. Baker having devised a new "weighting function" argument that runs its course in approximately 35 pages [1].

We conclude this discussion with a brief mention of the average case analysis that has been performed for the classical bin packing problem. The most extensive experimentation we know of was done in Johnson's thesis [37]; a summary of some of the results for uniform distributions of item sizes are given in Table 2. (Additional experiments are reported in [47], but these measure percentage of waste per bin rather than number of excess bins, and so are not readily comparable with our worst case results). Note that the ratios given are not strictly speaking averages of $R_A(L)$, since the value of OPT(L) could not be determined (its computation being an NP-complete problem). Instead these values are for the ratio of $A(L)$ to the sum of the item sizes (an obvious lower bound on OPT(L)). The interesting fact here is that the average behavior, although much better than the worst case behavior, still ranks the algorithms in the same relative order. Results for an approximation to a normal distribution, and for a distribution obtained by chopping up a set of items of size C into a random number of items, are slightly worse, but reflect the same trends [37]. It should not be expected, however, that average case ranking will always reflect worst case ranking. In particular, certain of the new algorithms specifically designed for improved worst case behavior (although possible not MFFD) may be comparatively bad on average.

Algorithm	UNIFORM (0,1.0)	UNIFORM (0,0.5)	UNIFORM (0,0.25)
NEXT FIT	31.1 [100.]	18.8 [100.]	7.4 [50.0]
NEXT-2 FIT	21.9 [85.0]	8.5 [50.0]	2.2 [25.0]
ALMOST WORST FIT	10.4 [70.0]	4.8 [50.0]	1.4 [25.0]
FIRST FIT	7.0 [70.0]	2.2 [50.0]	0.6 [25.0]
BEST FIT	5.6 [70.0]	2.2 [50.0]	0.5 [25.0]
GROUP FIT GROUPED	2.1 [50.0]	0.4 [33.3]	0.3 [20.0]
AWF DECREASING	2.0 [22.2]	0.2 [18.3]	0.2 [15.0]
FF DECREASING	1.9 [22.2]	0.1 [18.3]	0.2 [15.0]
BF DECREASING	1.9 [22.2]	0.1 [18.3]	0.2 [15.0]

Table 2. Percentages of excess bins required on the average in bin-packings of 25 lists with item sizes uniformly distributed within the stated ranges. [Percentage of excess in worst examples known are given in brackets].

It would be nice if we could resolve these average case questions theoretically, rather than experimentally. Unfortunately, it seems that currently available probabilistic techniques are only going to be useful for the simplest of algorithms. Most of the work to date has been on NEXT FIT. The first such analysis, due to Shapiro [49], was based on an approximation to the exponential distribution and estimated the expected value, given NF(L), of OPT(L). He concluded that as NF(L) approaches infinity, $R_{NF}(L)$ approaches 1 plus the average item size, when that average is C/5 or less.

The results for the uniform distribution are more exact. In [15], Coffman, Hofri, and So show that if item sizes are uniformly distributed between 0 and C, then the expected value $E(NF(L))$ is bounded by $(4/3)\cdot E(OPT(L)) + 4$. In addition, mathematical machinery is supplied for working out the expected values when the upper bound on the range of item sizes is less than C (note that the result for the range $[0,C]$ agrees quite well with the experimental results in Table 2).

One other result for items uniformly distributed in $[0,C]$ has recently appeared, and although this result is not so easy to extend to other ranges, it does concern a more complicated algorithm than NEXT FIT. In [20], Frederickson provides a rigorous proof of the rather intuitive fact that, for this particular distribution, the ratio $E(FFD(L))/E(OPT(L))$ approaches 1 as the number of items approaches infinity. Note that this ratio is not the same as $E(FFD(L)/OPT(L))$.

3. Variants with Different Ground Rules

In this section we survey results for variants on the classical one-dimensional bin packing problem in which the goal is still to minimize the number of bins used, but the ground rules for the packings are altered. Four basic modifications of the original problem are considered: (1) Packings in which bounds are placed in advance on the number of items that can be packed in a bin, (2) Packings in which a partial order is associated with the set of items to be packed and constrains the ways in which items may be packed, (3) Packings in which restrictions are placed on the items that may be packed in the same bin, and (4) Packings in which items may enter and leave the packing dynamically.

The first modification was considered by Krause, Shen, and Schwetman [44] as a model for multiprocessor scheduling under a single resource constraint when the number K of processors is fixed. In this case the items represent tasks to be executed, with the size of an item being the amount of the resource it requires (out of a total quantity of C). If we assume that all tasks have the same unit-length execution time, then a schedule corresponds to an assignment of tasks to integral starting times, such that at no time are there more than K tasks being executed or is more than C of the resource being used. The objective is to minimize the latest starting time. This corresponds to bin packing where the starting times represent bins, each of which can contain at most K items.

Krause et al. analyze three algorithms for this problem. The first two are just FIRST FIT and FIRST FIT DECREASING, suitably modified to take into account the bound on number of items per bin. The results are simply stated:

$$\frac{27}{10} - \left\lceil \frac{37}{10K} \right\rceil \le R_{FF}^{\infty} \le \frac{27}{10} - \frac{24}{10K} \quad ; \quad R_{FFD}^{\infty} = 2 - \frac{2}{K}$$

Note that as K goes to infinity, these bounds remain substantially worse than the corresponding bounds when the number of items per bin is not restricted (27/10 versus 17/10 and 2 versus 11/9). Thus the very existence of a limit, and not just its size, can have a substantial effect on the worst case behavior of the algorithms.

The third algorithm studied was alluded to in the previous section. ITERATED LOWEST FIT DECREASING uses a technique we shall be meeting again in the next section, so we shall describe it in detail. We first put the items in non-increasing order by size, as we do for FFD. We then pick some obvious lower bound q on OPT(L) and imagine we have q empty bins, $B_1, B_2, \ldots, B_q$. Place p_1 in B_1 and proceed through the list of items, packing p_i in a bin whose current contents has minimum total size (breaking ties by bin index, when necessary). If we ever reach a point where p_i does not fit in *any* of the q bins (either because the capacity C or the limit m is exceeded), we halt the iteration, increase q by 1, and start over. Eventually we will succeed in generating a packing for some value of q, and this is our output.

The running time of ILFD is $O(n^2 \log n)$, but this can be improved to $O(n \log^2 n)$ by using binary search on q. The performane bound proved for ILFD is $R_{ILFD}^{\infty} \leq 2$, which makes ILFD competitive with FFD. It is conjectured that the actual value of R_{ILFD}^{∞} is closer to the 4/3 value we cited in the last section for the case when there is no limit on the number of items per bin.

The second type of modification we wish to consider is the addition of a partial order $\lesssim$ on the set L of items. This arises in two potential applications of bin packing. One is again related to multiprocessor scheduling, and was studied by Graham, Yao, and ourselves in [24]. Suppose we have a set of unit-length tasks $p_1, \ldots, p_n$ with resource requirements subject to an over-all bound of C, as above, but with no limit on the number of processors. In this case a partial order $\lesssim$ is interpreted as follows: $p_i \lesssim p_j$ means that p_i must be executed *before* p_j, i. e., must be assigned to a bin with lower index than that to which p_j is assigned.

The other application in which partial orders arise is in "assembly line balancing," and is studied in Magazine and Wee [46]. Here the items represent tasks to be performed on a single product as it moves along an assembly line. Each is performed at one of a sequence of workstations B_1, B_2, etc., and the item sizes correspond to the times required to execute the tasks. The assembly line advances in discrete steps, stopping for a period of time C at each workstation. Thus a set of tasks can be assigned to a work station (bin) if their total time (size) does not exceed C. The goal is to minimize the number of workstations required. In this case a partial order $\lesssim$ has the following interpretation: $p_i \lesssim p_j$ means that in any assignment of tasks to workstations (bins), p_i must be performed before p_j (but they could be performed at the same workstation, merely by doing p_i before p_j within the total time C allowed, so this time p_i can go either in an earlier bin or the *same* bin as p_j).

Note that these two applications yield different interpretations of the partial order constraint within the bin packing context. See Figure 3. Although this difference might appear to be slight, its consequences, as shown in the figure, are nontrivial. The algorithm referred to there, ORDERED FIRST FIT DECREASING, is the best algorithm known for either version of the problem, but yields quite different guarantees. It is quite simple to describe. First, we order the items by non-increasing size, as with FFD. We then pack bins, rather than items, in sequence. Bin B_i is packed as follows: Place the largest unpacked item into B_i that the partial order will allow. Repeat until no more items can legally be packed into B_i.

Note that this algorithm can be applied to either version of the problem, so long as the partial order is interpreted appropriately. Note also that, in the absence of a partial order, this algorithm generates the same packing as FIRST FIT DECREASING and hence

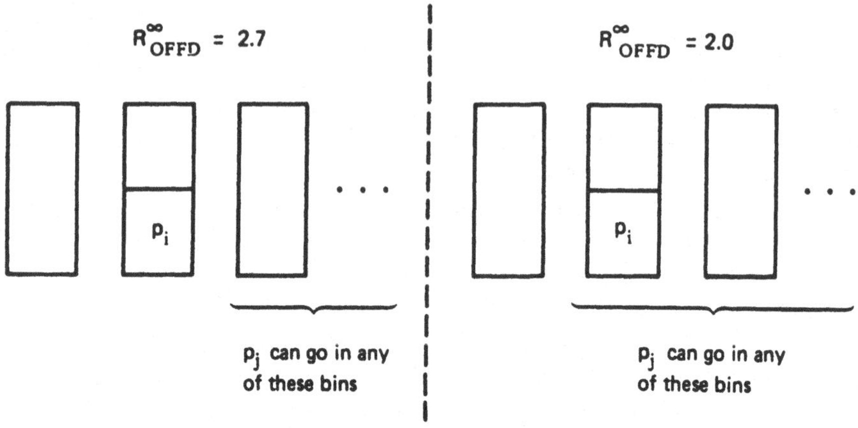

Figure 3. *Two interpretations of* $p_i \lesssim p_j$ and *their consequences.*

has an asymptotic worst case ratio of 11/9.

In the third of our four variants, a different type of restriction is placed on where an item may be packed. The basic idea is that only "closely related" items may go in a bin together. The one example we cite is from a paper by Chandra, Hirschberg, and Wong [11], although other potential applications of this type might come to mind. Here the items are thought of as having geographical locations. Putting them in the same bin corresponds to assigning them to a common facility (computing service, telephone switching center, etc.), where each such facility is assumed to have a standard capacity C. We desire that the items which are served by a common facility be in close proximity to each other, and this restricts the types of packings that we allow. The main results in [11] concern the case when the contents of a bin must all reside within the same unit square, although other figures, such as unit circles, are also considered. For the unit square case, a geometric algorithm is proposed and shown to have R_A^{∞} lying between 3.75 and 3.8.

The final variant we consider in this section is the "dynamic" bin packing problem, which models memory allocation in a computer. Here the bins correspond to "pages" in computer memory, and the items correspond to records which must be stored within the pages for certain specified periods of time. Associated with an item is thus not only a size $s(p_i)$, but also a beginning time $b(p_i)$ and an ending time $e(p_i)$. A packing is an assignment of items to bins such that at any time t and for any bin B, the items assigned to that bin which have begun by time t and not yet ended by that time have total size no greater than the capacity C.

We are currently studying this problem together with E. G. Coffman [13]. So far our research has concentrated on "on-line" algorithms, where in this case an on-line algorithm packs items in the order in which they begin, and may not use information about items which are to begin later, or the ending times for items which are currently in the packing (this lack of information mirrors the predicament often faced by actual computer storage allocators). We assume that once an item is assigned to a bin it cannot be moved to another bin, so that the items cannot be rearranged to take advantage of holes as they open up.

The algorithm FIRST FIT can be readily adapted to this situation, but the dynamic nature of the environment significantly impairs its performance. For the case when no item size exceeds $C/2$, we have $R_{FF}^{\infty}(1/2) = 1.5$ in the classical case, but in the dynamic case we show that *any* on-line algorithm must obey $R_A^{\infty}(1/2) \geq 5/3 = 1.666...$ For FIRST FIT we have been able to show by analytical means that R_{FF}^{∞} lies somewhere between 1.75 and 1.78. The case when items larger than $C/2$ are allowed is even more difficult to analyze (as seems usually to be the case with bin packing), but is clearly much worse. Here we know that R_{FF}^{∞} lies somewhere between 2.75 and 2.89, and *any* on-line algorithm must obey $R_A^{\infty} \geq 2.5$. Work is continuing on these and other questions.

4. Variants with Different Optimization Criteria

In this section we survey results for variants of the classical one-dimensional problem in which the objective is something other than minimizing the number of bins used. Most of these variants concern the case when the number of bins to be used is fixed.

For instance, consider the multiprocessor scheduling problem in which we are given m processors (bins), along with a set L of tasks (items) to be scheduled, i.e., assigned to processors. Each task has a specified length (item size) which represents the time required by a processor to execute it. Our goal is to assign tasks to processors so as to minimize the overall length of the schedule (the sum of the execution times for the busiest processor). Hence this is in a sense the "dual" of the classical bin packing problem: instead of being given L and C and asked to find a packing minimizing the number of bins m, we are given L and m and asked to find a packing which minimizes the required capacity C.

The initial work on approximation algorithms for this problem was done by R. L. Graham [31,32]. We can define worst case ratios as we did before, noting that in this case the asymptotic and absolute worst case ratios will coincide for reasonable algorithms – any worst case example can be converted to one with an arbitrarily large value for C merely by scaling up all the sizes by an appropriate multiplicative factor. (In those applications in which there is a fixed upper bound on the possible item sizes, asymptotic worst case bounds would make sense, but due to the nature of the problem we would tend to get $R_A^{\infty} = 1$ for most algorithms, e.g., see [41]). Thus we shall express results for this variant in terms of the absolute ratio R_A. Graham examined two basic algorithms for this problem. LOWEST FIT assigns the items to bins in order, placing p_i in a bin with current contents of minimum total size (ties broken by bin index when necessary). LOWEST FIT DECREASING (which we used as the iterated subroutine in the ILFD algorithm of the previous section) first sorts the items so that they are in non-increasing order by size and then applies LOWEST FIT. Fixing m, the number of bins, Graham was able to prove that $R_{LF} = 2 - (1/m)$ [31] and that $R_{LFD} = (4/3) - (1/3m)$ [32].

Although LOWEST FIT remains the currently best "on-line" algorithm known for this problems. LFD has since been dethroned as "off-line" champ. Sahni [48] has shown that for any fixed value of m and any $\epsilon > 0$ there exists a polynomial time algorithm with $R_A = 1+\epsilon$. Unfortunately, these algorithms are exponential in m (and polynomial in $1/\epsilon$), and are not terribly attractive for $m > 3$ or for very small ϵ.

An algorithm which is polynomial in m and still beats LFD was presented by E. G. Coffman and ourselves in [12]. The algorithm works on an iterative principle much like that of ILFD. Called MULTIFIT DECREASING, the algorithm works by guessing a capacity C and then applying FFD to the list. The next guess is either larger or smaller, depending on whether FFD used more than m bins of that capacity to pack the list or not. By using an appropriate binary search strategy and limiting the number of iterations performed to some small number k, we obtain an algorithm, denoted MF(k), which obeys the bound $R_{MF[k]} \le 1.220+(1/2)^k$, independently of the value of m. This beats the LFD bound for all $m>2$ when $k\ge6$, at the cost of only a small increase in running time. The worst behavior known for this algorithm is shown in examples constructed by Friesen [21], which imply that $R_{MF[k]} \ge 13/11=1.18181...$ for $m\ge13$. (Better upper bounds are known for the cases when $m\le7$ [12]).

We remark in passing that the results for MULTIFIT DECREASING are proved by considering the following bin packing variant: Suppose we are given two sets of bins, one with bins all of capacity α, the other with bins all of capacity β. What is the asymptotic worst case ratio of the number of β-capacity bins used by FF (or FFD) to the minimum number of α-capacity bins needed? This question, first raised in [23], is investigated in detail in Friesen's thesis [21] as well as in [12], and is used in [23] for proposing a conjectural explanation of the mysterious fraction 17/10 in the original theorem about R_{FF}^{∞} (for more on this conjecture, see Section 7).

A second bin packing variant based on a fixed number of bins was proposed by Chandra and Wong [10] as a model for arm contention in disk pack computer storage. Item sizes represent access probabilities, and if two items from the same bin are requested at the same time, this results in contention, which we wish to minimize. If there are m bins and the total sizes of their contents are $S_1, S_2, \ldots, S_m$, then an estimate of the total contention is given by $\sum_{i=1}^{m}(S_i)^2$. The goal is to minimize this quantity (which can roughly be accomplished by making the bin totals all as close to each other as possible). The algorithm LFD is analyzed in this context, and it is shown that for this problem $1.0278 \le R_{LFD} \le 1.0417$. In [18], Easton and Wong consider the variant on this variant in which no bin can contain more than K items. They analyze an appropriately modified version of LFD and show that $R_{LFD} \le 4/3$.

Wong and Yao [52] consider yet another variant, based on a storage allocation problem where the goal is to minimize access time [54]. Suppose we wished to *maximize*, rather than minimize, the sum of squares in the above case where K items per bin are allowed. This might be considered as a bin packing problem where all items have both a size $s(p)$ = 1 and an arbitrary weight $w(p)$, the bin capacity is K, and the goal is to pack the items into m bins so as to maximize the sum of the squares of the total weight in each bin. As observed in [54], this is of course an easy matter: merely put the K largest items in the first bin, the next K largest in the second bin, etc. Wong and Yao consider the generalization where the sizes as well as the weights are arbitrary.

In order that results for this maximization problem can be compared directly to those for the minimization problems we have been studying so far, we shall define $R_A(L)$ to be OPT(L)/A(L) for any approximation algorithm A (this is the inverse of our definition for minimization problems). R_A and R_A^{∞} are then defined as before and lie in the range $[1,\infty]$. Wong and Yao propose a heuristic based on ordering the items by non-increasing density (weight divided by size) and then applying NEXT FIT. They show that this heuristic satisfies $R_A \leq 2$.

The final variant we consider in this section is again a maximization problem that fixes the number of bins and the bin capacity. This time the goal is to pack as *many* items as possible into the bins. Coffman, Leung, and Ting [17] consider the algorithm FIRST FIT INCREASING, which first sorts the items into non-decreasing order by size, and then applies FF until an item is reached which will not fit in any of the bins (which implies that none of the remaining items will fit either). They show that $R_{FFI} = 4/3$. In [16], Coffman and Leung consider an algorithm that, like ILFD and MFD, involves iteration. Their algorithm, denoted FFD*, works as follows: First sort the items in non-increasing order by size, and then apply FF. If some item fails to fit, stop, delete the first (largest) item in the list, and reapply FF to the shorter list. Repeat this until a list is obtained that FF *does* pack into the *m* bins. Coffman and Leung show that FFD* will always pack at least as many items as FFI, and indeed obeys the better bounds $8/7 \leq R^{\infty}_{FFD^*} \leq 7/6$, making the added complexity of FFD* over FFI worth the effort.

5. Vector Packing

In this section we consider one way of generalizing the classical one-dimensional bin packing problem to higher dimensions. Instead of each $s(p)$ being a single number, we consider the case when it is a d-dimensional vector $s(p) = \langle s_1(p), s_2(p), \ldots, s_d(p) \rangle$. The bin capacity is also a d-dimensional vector $\langle C, C, \ldots, C \rangle$, and the goal is to pack the items in a minimum number of bins, given that the contents of any given bin must have vector sum less than or equal to its capacity. This problem models multiprocessor scheduling of unit-length tasks in the case when there are d resources, rather than just one as we assumed before. For simplicity we have normalized the amounts of resources available so that all d bounds are the same.

Note that the two-dimensional version of this problem is not the same as the problem of packing rectangles (to be discussed in the next section). A vector $\langle s_1(p), s_2(p) \rangle$ could be thought of as representing a rectangle with length $s_1(p)$ and width $s_2(p)$, and a bin of capacity $\langle C, C \rangle$ as a square into which the rectangles are to be packed. However, the only types of packings allowed here would correspond to ones in which the rectangles were placed corner to corner, diagonally across the bin, as in Figure 4.

In [43], Kou and Markowsky show that any "reasonable" algorithm, i.e., one which does not yield packings in which the contents of two non-empty bins can be combined into a single bin, obeys the bound $R_A \leq d+1$, where d is the number of dimensions (an alternative proof can be found in [22], although there the theorem is not stated in its full generality). We note in passing that, in spite of the obvious desirability of the above "reasonable" property, not all the algorithms we have mentioned so far are reasonable – an obvious offender is NEXT FIT. However, FIRST FIT, FIRST FIT DECREASING, and many of our favorites *are* reasonable and hence do obey the above-mentioned, not very impressive (when k is large), bound. They are, in fact, better than reasonable, but not by as much as one would like. In [24], Garey, Graham, Johnson, and Yao analyze the d-dimensional problem and appropriate adaptations of FF and FFD to this multi-dimensional case (in FFD the items are sorted in non-increasing order by the maximum components of their size vectors). They show that $R^{\infty}_{FF} = d + 7/10$, which reduces to the familiar 17/10

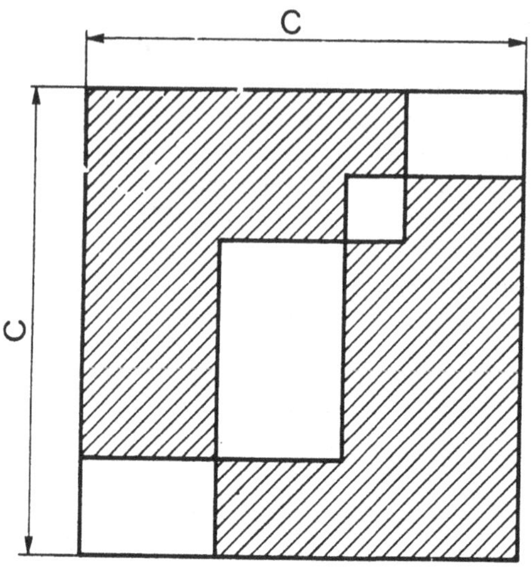

Figure 4. *Two–dimensional vector packing.*

result in the one-dimensional case, and that $d \leq R_{FFD}^{\infty} \leq d + 1/3$. To date, no one has found any polynomial time approximation algorithm for the general d-dimensional problem with $R_A^{\infty} < d$. Yao has shown [53] that any algorithm that is *faster* than FF or FFD, i.e., that has a running time that is $o(n\log n)$ in the decision tree model of computation, *must* have $R_A^{\infty} \geq d$. This is a rather unpromising state of affairs, although, as indicated by extensive simulation results of Maruyama, Chang, and Tang for FF, FFD, and a variety of other algorithms [47], average case behavior may not be nearly as bad here as the worst case bounds.

In the variant on this problem in which a partial order is present, however, things definitely get worse. Suppose that the set of items has a partial order $\lesssim$ associated with it that constrains the allowable packings as in Section 3 (the multiprocessor scheduling rather than the assembly line balancing case). In this case the natural generalization of the ORDERED FIRST FIT DECREASING algorithm of Section 3 can be shown [24] to obey $(1.691)d+1 \leq R_{OFFD}^{\infty} \leq (1.7)d+1$, a definite step backwards from our bounds when no partial order was present (the result mentioned in Section 3 for the one-dimensional version of this problem is a special case of this result). Similar results are obtained for the algorithm ORDERED FIRST FIT BY LEVEL, which works the same way as OFFD, except that instead of ordering items by non-increasing maximum size component, they are ordered by non-increasing "level" in the partial order [24]: $R_{OFFL}^{\infty} = (1.7)d+1$. That some type of pre-ordering is necessary for even this standard of performance follows from the fact that the algorithm without any pre-ordering, ORDERED FIRST FIT, has $R_{OFF}^{\infty} = \infty$ [24].

6. Rectangle Packing

In this section we consider one of the currently most active areas of bin packing research: the problem of packing rectangles into two-dimensional bins. The first version of this problem to be studied from a performance guarantee point of view is due to Baker, Coffman, and Rivest [5], and can be thought of as modelling a variety of problems, from computer scheduling to stock cutting. In this version, the items p_i are rectangles, with height h_i and width w_i. The goal is to pack them in a vertical strip of width C, so as to minimize the total height of the strip needed. The rectangles must be packed orthogonally, that is, no rotations are allowed: all rectangles must have their width parallel to the bottom of the strip.

The orthogonality restriction is justified on the basis of the proposed application to scheduling. Here the items once again correspond to tasks. The height of an item is the amount of processing time it requires, and its width is the amount of contiguous memory it needs. The strip width C is then the total amount of memory available; the strip length is the amount of time needed to schedule all the items. In this application it makes no sense to rotate a rectangle, even by ninety degrees, since execution time is not in general directly translatable into memory requirement (despite all the recent theoretical work on time-space trade-offs).

The application to stock-cutting is as follows: In a variety of industrial settings, the "raw" material involved comes in rolls, for instance, rolls of paper, rolls of cloth, rolls of sheet metal, etc. From these rolls we may wish to cut patterns (for labels, clothes, boxes, etc.) or merely just shorter, narrower rolls. In the simplest case, we can view the objects we wish to cut from the rolls as being, or approximating, rectangles. We minimize our wastage if we minimize the amount of the roll (the strip length) used. Once again some form of orthogonality may be justified, since in many applications, the cutting is done by blades that must be either parallel or perpendicular to the strip, and the material may have a bias that dictates the orientation of the rectangles. However, ninety degree rotations may in some cases be allowable, and we will later say a bit about how the results we discuss can be extended to take this into account.

In [5], Baker, Coffman, and Rivest consider a variety of strip packing algorithms based on a "bottom up – left justified" (BOTTOM-LEFT for short) packing rule. In a BOTTOM-LEFT packing, items are packed in turn, each item being placed as near to the bottom of the strip as it will fit and then as far to the left as it can be placed at that bottom-most level. Note that there is a difference in kind between two-dimensional packing rules, such as the BOTTOM-LEFT rule, and one-dimensional rules such as FIRST FIT and NEXT FIT. In the one-dimensional case there always exists an ordering of the items such that FIRST FIT (NEXT FIT) constructs an optimal packing. However, this is not the case for BOTTOM-LEFT. In fact, D. J. Brown has constructed instances in which the best BOTTOM-LEFT packing possible still yields a strip whose height is 5/4 times optimal [9].

However, although no preordering of the items may be able to yield an optimal packing, some may still be better than others. Various BOTTOM-LEFT algorithms can be considered, depending on how (if at all) the set of rectangles is initially preordered. It turns out that only one of the standard orderings seems to make a difference as far as worst case behavior is concerned. If we let BL stand for the unadorned BOTTOM-LEFT algorithm, and BLIW, BLIH, BLDW, and BLDH stand for the algorithm with preordering by increasing (i.e., non-decreasing) width, increasing height, decreasing width, and decreasing height, then we have

$$R_{BL} = R_{BLIW} = R_{BLIH} = R_{BLDH} = \infty \quad ; \quad R_{BLDW} = 3$$

For the special case of squares ($h_i = w_i$ for all p_i), the BLDH algorithm becomes equivalent to BLDW, and the result improves to $R_{BLDW} = 2$ [5].

For the case of arbitrary rectangles, subsequent work has yielded some improvements. The FIRST FIT DECREASING HEIGHT "level" algorithm of Coffman, Garey, Johnson, and Tarjan [14] (to be described later) can be shown to have $R_{FFDH} = 2.7$, and a recent algorithm of Sleator [50] is the current champ with $R_A = 2.5$.

These results all concern *absolute* worst case performance ratios. Indeed, for this problem it would again seem as if absolute and asymptotic performance ratios should be equivalent, since heights can be scaled to arbitrarily large values. However, such scaling may not be sensible in many practical applications, where some strict upper bound on height may be imposed. In this case, asymptotic analysis may be a more meaningful measure, giving us guarantees that hold as the optimal strip length becomes very large with respect to this maximum possible item height. As might be expected, these asymptotic guarantees can be better than the absolute ones (although they do not equal 1, as they would if all rectangle widths were equal, thus reducing us to the capacity minimization problem of Section 3). For instance, $R_{BLDW}^{\infty} = 2$, an improvement of 1 over the absolute guarantee for BLDW, but a long way from optimal.

The search for strip packing algorithms with better asymptotic worst case ratios was taken up in [14] by Coffman, Tarjan, and ourselves. The new algorithms were based on a different type of packing rule, suggested by Golan [30], and were called "level" algorithms. These algorithms involve an attempt to apply our old knowledge about one-dimensional bin packing. Note that if all rectangles have the same height, the two-dimensional problem essentially reduces to the one-dimensional case: in an optimal packing the items may be placed in rows or "levels." Each level in the packing then corresponds to a bin and the height of the packing corresponds to the number of bins used. The basic idea of a level algorithm is the following: First, the items are preordered by non-increasing height. The packing is then constructed as a sequence of *levels*, each rectangle being placed so that its bottom rests on one of these levels. The first level is simply the bottom of the bin. Each subsequent level is defined by a horizontal line drawn though the top of the tallest rectangle on the previous level. This is best illustrated by considering the two basic level algorithms proposed in [14].

In the algorithm NEXT FIT DECREASING HEIGHT rectangles are packed left-justified on a level until the next rectangle will not fit, in which case it is used to start a new level above the previous one, on which the packing proceeds. Note the analogy with the one-dimensional NEXT FIT algorithm. In the FIRST FIT DECREASING HEIGHT algorithm (another analog), each rectangle is placed left-justified on the first (i.e., lowest) level in which it will fit. If none of the current levels has room, a new one is started as with the NFDH algorithm. See Figure 5 for an example of an FFDH packing.

At first glance, one would expect NFDH and FFDH to be worse than their one-dimensional counterparts, given all the space that may be wasted in a level above items which are shorter than the first one. However, it turns out that this wasted space is strictly bounded, and by a collapsing sum argument it can be concluded that, exactly as in the one-dimensional case,

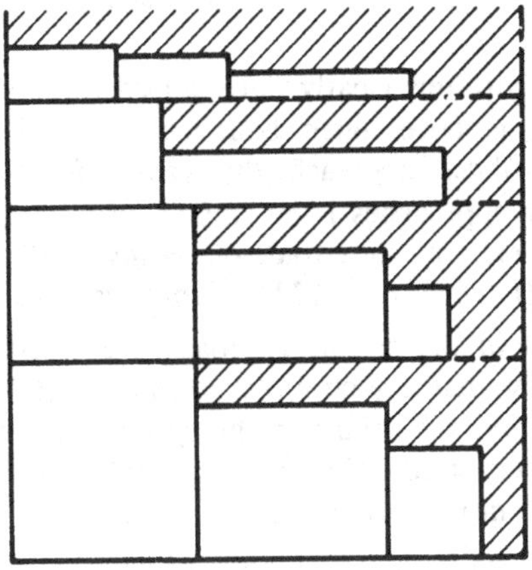

Figure 5. *An example of an FFDH packing.*

$$R^{\infty}_{NFDH} = 2 \quad ; \quad R^{\infty}_{FFDH} = 1.7$$

The results for bounded item widths also mimic their one-dimensional counterparts, and for the special case of squares, the asymptotic worst case ratio is reduced to 1.5.

Having beaten the BOTTOM LEFT algorithms with FFDH, the question next arises, can we beat FFDH? In the one-dimensional case, FF is beaten by FFD. However, FFD requires that the items be preordered by non-increasing size, which here corresponds to non-increasing width, and since FFDH already requires items to be preordered by height, any additional preordering becomes impossible. Fortunately, there are ways of approximating FFD in the two-dimensional case. In the SPLIT FIT algorithm of [14], the set of rectangles is partitioned into two parts, those with width exceeding $C/2$ and those without, and each subset is ordered by non-increasing height. Packings for the two sets are then combined in an involved manner, and the result is an algorithm with $R^{\infty}_{SF} = 1.5$. This idea of splitting the set of rectangles into subsets according to width can be carried even further, and Baker, Brown, and Katseff now have devised a much more complicated algorithm [3] for which $R^{\infty}_{A} \leq 5/4$, a bound which is very close to the 11/9 guarantee provided by FFD in the one-dimensional case.

So far, all the rectangle packing algorithms we have discussed for which $R^{\infty}_{A} \leq \infty$ have involved some preordering of the rectangles, and hence are not "on-line" algorithms. However, such algorithms might well be required in scheduling applications, and so the question of finding an on-line algorithm with reasonable worst case behavior becomes relavant. Baker and Schwartz, in [6], show that such algorithms exist by devising what they call "shelf" algorithms. These are variants on the level algorithms above in which levels, rather than being determined by their tallest item, come in fixed sizes. If we assume that 1 is an *a priori* upper bound on rectangle height, the standard levels will come in heights r^{-k}, $k \geq 0$, for some prespecified value of r, $0<r<1$. Whenever a rectangle p_i is to be packed in the NEXT FIT SHELF(r) algorithm, one first determines that value of k such that $r^{k+1} < h_i \leq r^k$. If there is a level of height r^k already in the packing, and p_i will fit in the currently active one, it is placed there. Otherwise it is placed in a new such level, which becomes the currently active one for that height. The algorithm FIRST FIT SHELF(r) is defined analogously.

Although these shelf algorithms clearly have considerable space-wasting potential, it turns out that the wastage is once again boundable, and in fact $R^{\infty}_{NFS(r)} = 2/r$ and $R^{\infty}_{FFS(r)} = 1.7/r$. Note that these approach the values for NFDH and FFDH as r approaches 1. However, as r approaches 1 the amount of wastage to be expected in small examples increases as $C/(1-r)$, and so a trade-off is involved. The best absolute worst case ratio is obtained by FFS(r) when $r \sim .622$, in which case we have $R_{FFS(r)} \sim 6.9863$.

The limitations inherent in the on-line approach are investigated by Baker, Brown, and Katseff in [2]. They show that *any* on-line algorithm must obey $R_A \geq 2$. (The paper also contains bounds for on-line algorithms in the special case where they happen to be given sets of rectangles in some sorted order, but must still pack each item in turn, without being able to look ahead or to move an item once it is placed).

There have been two papers to date that cover average case analysis for strip packing. In [20], Frederickson proposes an off-line algorithm specifically designed for the case when item sizes and widths are independently and uniformly distributed between 0 and C, and combining FFD with specially tuned shelf sizes. Although the expected wastage may be large in absolute terms (proportional to $n^{3/4}$), the ratio of expected strip length to a lower bound on the optimal length (obtained by dividing the expected total area of rectangles by the strip width C) approaches 1 as n goes to infinity.

In [36], Hofri concentrates more on the on-line case, extending his earlier work with Coffman and So [15] on the expected behavior of one-dimensional NEXT FIT to the strip packing problem. He considers two new on-line algorithms. The first is a level algorithm in which there is no initial reordering of the list of items, and hence the height of a level is not determined by its first rectangle but by the tallest, whichever one that might be. Otherwise the packing rule is basically a NEXT FIT one: an item is packed in the current level unless it won't fit along the bottom, in which case it starts a new level, whose bottom is coincident with the top of the tallest item in the earlier level. Hofri calls this algorithm NEXT FIT, as opposed to NEXT FIT DECREASING HEIGHT where the items are preordered. Hofri's other new on-line algorithm is appropriate in the case where ninety degree rotations are allowed, and is called ROTATABLE NEXT FIT. This algorithm is the same as NEXT FIT except that each item is rotated before it is packed so that its height does not exceed its width.

Both of these two new on-line algorithms have $R^{\infty}_A = \infty$ and so are not very attractive from a worst case point of view. However, Hofri shows that when heighths and widths are independent and uniformly distributed between 0 and C, they are not that much worse than NEXT FIT DECREASING HEIGHT, which has $R^{\infty}_{NFDH} = 2$ and is not an on-line algorithm. As the number of items goes to infinity, Hofri's results indicate that NEXT FIT DECREASING HEIGHT averages roughly 4/3 times the above-mentioned lower bound on optimal strip length. ROTATABLE NEXT FIT is only slightly worse, and NEXT FIT's ratio is only about 3/2.

Having introduced the case where ninety degree rotations are allowed, we should mention that some of the worst case results mentioned above also apply to this case, in that the values of R_A and R^{∞}_A are unchanged if such rotations are allowed in the construction of optimal packings. This holds true in particular for NFDH and BLDW, since the proofs of the bounds for these algorithms are based on pure area arguments. So far no algorithm has been found that attains improved guarantees by actually using such rotations itself, and the results mentioned above for the performance of strip packing algorithms when all items are squares (and hence ninety degree rotation cannot help) indicate that we can expect only limited improvements.

A rectangle packing problem closely related to strip packing is that of packing a given set of rectangles into an enclosing rectangle of minimum area. Strip packing is the special case where the width is fixed. In this general problem both length and width are allowed to vary. To date there has not been much work on this problem from a performance guarantee point of view. Two papers of interest have addressed the case when all the items to be packed are squares. In [42], Kleitman and Krieger show that a collection of squares whose total area is unity can always be packed into a rectangle with area $4/\sqrt{6}$, and this is the minimum area for which such a packing is guaranteed. Furthermore, a $2/\sqrt{3}$ by $\sqrt{2}$ rectangle is the unique rectangle that will always suffice. In [19], Erdös and Graham consider the minimum sized square required to contain a collection of unit squares, and show that this size can be nontrivially decreased if rotations other than ninety degrees are allowed.

The final rectangle packing problem we shall consider is a much more straightforward generalization of the one-dimensional case than was strip packing. Here the problem is once again simply to minimize the number of bins used, the bins now being large rectangles of some fixed dimensions into which the given set of rectangles must be packed. We first note that if the number of possible rectangle sizes is sufficiently small, a Gilmore-Gomory style linear programming approach can be applied [51] with useful results. For the general problem, the only algorithm which to date has been analyzed from the worst case point of view is a composite algorithm we have proposed ourselves. We shall denote this algorithm by "FFDH * FFD," as it is based on the algorithms FFDH for strip packing and FFD for one-dimensional bin packing. The idea of the algorithm goes as follows: Suppose our standard bin has width W and height H. First use the FFDH algorithm to pack the set of rectangles into a strip of width W. Next, decompose this packing into blocks corresponding to the levels created by FFDH. Each block can be viewed as a rectangle of width W and height the height of the level. Thus packing these blocks into rectangular bins of width W becomes a simple one-dimensional bin packing problem, where the size of an item (block) is its height. Apply FFD to this one-dimensional problem.

The analysis of this algorithm is not yet complete [27], but preliminary results show that $1.9666... \le R^{\infty}_{FFDH*FFD} \le 2.2666....$ Note that this leaves open the possibility that $R^{\infty}_{FFDH*FFD} = (R^{\infty}_{FFDH})(R^{\infty}_{FFD}) = (17/10)(11/9) = 2.0777...$, although the proof of such a result might well represent quite a challenge, especially if the product rule also carries over into proof lengths.

7. Directions for Future Research

In this section we briefly mention some of the open problems that have hounded us during our long association with the packing of bins. First there is the basic problem of finding simpler and more general proof techniques. Although we have concentrated here on the results rather than the proof techniques, it is the case that most of the results we have cited have only been proved by very problem-specific techniques, with far to much ad hoc-ary and detailed case analysis for the mathematically pure at heart. To be sure, researchers have been able to use intuition gained in studying the classical one-dimensional case in deriving results for the more complicated variants and generalizations, but unfortunately this is not often very apparent in the resulting proofs. The closest we have come to a general method for proving results of this sort is the "weighting function" approach, as illustrated by [12], [16], [24], and [39], but so far the details of how this approach is used vary considerably from one problem to the next. Any progress toward finding ways to shorten proofs of similar results in the future would be heartily cheered by those working in the area.

On a less fundamental level, there is of course the problem of finding better algorithms for the various problems, especially in the area of rectangle packing, and of tightening up the bounds on the algorithms already proposed but incompletely analyzed (we ourselves are still working on the MODIFIED FIRST FIT DECREASING algorithm from Section 1 and the FFDH $*$ FFD algorithm from Section 6).

There is also always room for new bin packing variants, the key being to find a variant which models practical problems *and* is susceptible to meaningful analysis. For instance, we are often asked about the case when there are different types of bins (i.e., different sizes, different costs, etc.). Can this be modeled in such a way that the results proven for it are not hopelessly narrow special cases?

Finally, we would like to mention two technical problems for which we would particularly like to see solutions. The first concerns a conjecture of Graham and ourselves mentioned in Section 4. When people hear about these bin packing results, their first reaction is often to wonder out loud where such funny rational numbers as 17/10 and 11/9 "come from." At least in the case of 17/10, there appears to be a number-theoretic explanation. If α is a positive rational number, let a *partition* of α be any sequence $\bar{n} = (n_1, n_2, \cdots)$ of integers greater than one, at least two of which do not equal 2, such that $\sum 1/n_i = \alpha$. The *weight* of a partition $\bar{n}$ is then defined to be $W(\bar{n}) = \sum 1/(n_i - 1)$. It turns out that 17/10 is the maximum, over all partitions $\bar{n}$ of $\alpha = 1$, of the quantity $W(\bar{n})$. We conjecture that this is not a fluke. For any $\alpha > 0$, define

$$W(\alpha) = \max\{W(\bar{n}): \bar{n} \textit{ is a partition of } \alpha\}$$

If $R_{FF}^{\infty}[\alpha]$ is defined to be the asymptotic worst case ratio for FIRST FIT when FIRST FIT uses bins of size 1 and the optimal packing uses bins of size α, we conjecture that $R_{FF}^{\infty}[\alpha] = W(\alpha)$. (This is discussed in greater detail in [23], where we also provide proofs for a few special cases).

The other one of our favorite problems is a very fundamental one about lower bounds. In Sections 2, 3, and 6 we mentioned lower bounds on R_A which hold for all on-line algorithms, and in Section 5 we mentioned an analogous result proved by Yao for vector packing algorithms that run in time $o(n \log n)$ in a certain model of computation. However, even though the best polynomial time approximation algorithm we know has $R_A^{\infty} \geq 1.18333$, no non-trivial bound (i.e., a bound stronger than the obvious $R_A \geq 1$) is known to hold for all polynomial time approximation algorithms, even if we assume that NP-complete problems cannot be solved by polynomial time algorithms (i.e., "P does not equal NP" in the jargon of complexity theorists). In fact, for all we know there could be a polynomial time algorithm for the one-dimensional problem that always comes within one bin of the optimal number. The NP-completeness proofs for bin packing do not rule this out. We doubt that such algorithms do exist, however, and so suspect that some result should be provable of the form "If there is a constant c and a polynomial time algorithm A such that $A(L) \leq OPT(L) + c$ for all instances L, then P equals NP." Results like this, and even stronger ones, have been proved for other problems (see [26]), but it seems very hard to prove such results for bin packing, and we would consider any progress in this direction to be quite worthwhile.

REFERENCES

1. Baker, B. S., private communication (1979).
2. Baker, B. S., Brown, D. J., and Katseff, H. P., "Lower bounds for on-line two-dimensional packing algorithms," *Proc. 1979 Conf. on Information Sci. and Systems*. Dept. of Electrical Eng., Johns Hopkins University, Baltimore, MD (1979), 174-179.
3. Baker, B. S., Brown, D. J., and Katseff, H. P., "A 5/4 algorithm for two-dimensional packing," (to appear).
4. Baker, B. S. and Coffman, E. G., Jr., "A tight asymptotic bound for next-fit-decreasing bin packing," (to appear).
5. Baker, B. S., Coffman, E. G., Jr., and Rivest, R. L., "Orthogonal packings in two dimensions," *SIAM J. Comput.* (to appear).
6. Baker, B. S. and Schwarz, J. S., "Shelf algorithms for two-dimensional packing problems," *Proc. 1979 Conf. on Information Sci. and Systems*, Dept. of Electrical Eng., Johns Hopkins University, Baltimore, MD (1979), 273-276.
7. Brown, A. R., *Optimum Packing and Depletion*, American Elsevier, New York (1971).
8. Brown, D. J., "A lower bound for on-line one-dimensional bin packing algorithms," Technical Report R-864 (1979), Coordinated Science Laboratory, University of Illinois, Urbana, IL.
9. Brown, D. J., private communication (1980).
10. Chandra, A. K. and Wong, C. K., "Worst-case analysis of a placement algorithm related to storage allocation," *SIAM J. Comput.* **4** (1975), 249-263.
11. Chandra, A. K., Hirschberg, D. S., and Wong, C. K., "Bin packing with geometric constraints in computer network design," Computer Science Research Report RC 6895 (1977), IBM Research Center, Yorktown Heights, New York.
12. Coffman, E. G., Jr., Garey, M. R., and Johnson, D. S., "An application of bin-packing to multiprocessor scheduling," *SIAM J. Comput.* **7** (1978), 1-17.
13. Coffman, E. G., Jr, Garey, M. R., and Johnson, D. S., "Dynamic bin packing," (to appear).
14. Coffman, E. G., Jr., Garey, M. R., Johnson, D. S., and Tarjan, R. E., "Performance bounds for level-oriented two-dimensional packing algorithms," *SIAM J. Comput.* (to appear).

15. Coffman, E. G., Jr., Hofri, M., and So, K., "A stochastic model of bin packing," (to appear).

16. Coffman, E. G., Jr., and Leung, J. Y., "Combinatorial analysis of an efficient algorithm for processor and storage allocation," *SIAM J. Comput.* **8** (1979), 202-217.

17. Coffman, E. G., Jr., Leung, J. Y., and Ting, D. W., "Bin packing: maximizing the number of pieces packed," *Acta Informatica* **9** (1978), 263-271.

18. Easton, M. C. and Wong, C. K., "The effect of a capacity constraint on the minimal cost of a partition," *J. Assoc. Comput. Mach.* **22** (1975), 441-449.

19. Erdös, P. and Graham, R. L., "On packing squares with equal squares," *J. Combinatorial Theory Ser. A* **19** (1975), 119-123.

20. Frederickson, G. N., "Probabilistic analysis for simple one- and two-dimensional bin packing algorithms," (to appear).

21. Friesen, D. K., "Sensitivity analysis for heuristic algorithms," Technical Report UIUCDCS-R-78-939 (1978), Dept. Comp. Sci., Univ. of Illinois, Urbana, IL.

22. Garey, M. R. and Graham, R. L., "Bounds on multiprocessor scheduling with resource constraints," *SIAM J. Comput.* **4** (1974), 187-200.

23. Garey, M. R., Graham, R. L., and Johnson, D. S., "On a number-theoretic bin packing conjecture," *Proc. 5th Hungarian Combinatorics Colloquium*, North-Holland, Amsterdam (1978), 377-392.

24. Garey, M. R., Graham, R. L., Johnson, D. S., and Yao, A. C., "Resource constrained scheduling as generalized bin packing," *J. Combinatorial Theory Ser. A* **21** (1976), 257-298.

25. Garey, M. R. and Johnson, D. S., "Approximation algorithms for combinatorial problems: an annotated bibliography," in J. F. Traub (ed.), *Algorithms and Complexity: New Directions and Recent Results*, Academic Press, New York (1976), 41-52.

26. Garey, M. R. and Johnson, D. S., *Computers and Intractability: A Guide to the Theory of NP-Completeness*, W. H. Freeman and Co., San Francisco (1979).

27. Garey, M. R. and Johnson, D. S., in preparation.

28. Gilmore, P. C. and Gomory, R. E., "A linear programming approach to the cutting stock problem," *Operations Res.* **9** (1961), 849-859.

29. Gilmore, P. C. and Gomory, R. E., "A linear programming approach to the cutting stock problem - Part II," *Operations Res.* **11** (1963), 863-888.

30. Golan, I., "Orthogonal oriented algorithms for packing in two dimension," Draft (1978).

31. Graham, R. L., "Bounds for certain multiprocessing anomalies," *Bell System Tech. J.* **45** (1966), 1563-1581.

32. Graham, R. L., "Bounds on multiprocessing timing anomalies," *SIAM J. Appl. Math.* **17** (1969), 263-269.

33. Graham, R. L., "Bounds on multiprocessing anomalies and related packing algorithms," *Proc. 1972 Spring Joint Computer Conference*, AFIPS Press, Montvale, NJ (1972), 205-217.

34. Graham, R. L., "Bounds on performance of scheduling algorithms," in E. G. Coffman, Jr. (ed.), *Computer and Job-Shop Scheduling Theory*, John Wiley & Sons, New York (1976), 165-227.

35. Graham, R. L., Lawler, E. L., Lenstra, J. K., and Rinnooy Kan, A. H. G., "Optimization and approximation in deterministic sequencing and scheduling: a survey," *Annals Disc. Math.* **5** (1979), 287-326.

36. Hofri, M., "1.5 dimensional packing: expected performance of simple level algorithms," Technical Report No. 147 (1979), Dept. Computer Science, Technion, Haifa, Israel.

37. Johnson, D. S., "Near-optimal bin packing algorithms," Technical Report MAC TR-109 (1973), Project MAC, Massachusetts Institute of Technology, Cambridge, Mass.

38. Johnson, D. S., "Fast algorithms for bin packing," *J. Comput. Syst. Sci.* **8** (1974), 272-314.

39. Johnson, D. S., Demers, A., Ullman, J. D., Garey, M. R., and Graham, R. L., "Worst-case performance bounds for simple one-dimensional packing algorithms," *SIAM J. Comput.* **3** (1974), 299-325.

40. Karp, R. M., "Reducibility among combinatorial problems," in R. E. Miller and J. W. Thatcher (ed.), *Complexity of Computer Computations*, Plenum Press, New York (1972), 85-103.

41. Kaufman, M. T., "An almost-optimal algorithm for the assembly line scheduling problem," *IEEE Trans. Computers* **C-23** (1974), 1169-1174.

42. Kleitman, D. J. and Krieger, M. K., "An optimal bound for two dimensional bin packing," *Proc. 16th Ann. Symp. on Foundations of Computer Science*, IEEE Computer Society, Long Beach, CA (1975), 163-168.

43. Kou, L. T. and Markowsky, G., "Multidimensional bin packing algorithms," *IBM J. Res. & Dev.* **21** (1977), 443-448.

44. Krause, K. L., Shen, Y. Y., and Schwetman, H. D., "Analysis of several task-scheduling algorithms for a model of multiprogramming computer systems," *J. Assoc. Comput. Mach.* **22** (1975), 522-550.

45. Liang, F. M., "A lower bound for on-line bin packing," *Information Processing Lett.* **10** (1980), 76-79.

46. Magazine, M. J. and Wee, T. S., "The generalization of bin-packing heuristics to the line balancing problem," Working Paper No. 128 (1979), Dept. Mgmt. Sci., University of Waterloo, Waterloo, Ontario.

47. Maruyama, K., Chang, S. K., and Tang, D. T., "A general packing algorithm for multidimensional resource requirements," *Internat. J. Comput. Infor. Sci.* **6** (1977), 131-149.

48. Sahni, S., "Algorithms for scheduling independent tasks," *J. Assoc. Comp. Mach.* **23** (1976), 116-127.

49. Shapiro, S. D., "Performance of heuristic bin packing algorithms with segments of random length," *Information and Control* **35** (1977), 146-148.

50. Sleator, D. K. D. B., "A 2.5 times optimal algorithm for bin packing in two dimensions," *Information Processing Lett.* **10** (1980), 37-40.

51. Taylor, D. B., "Container stacking: an application of mathematical programming," Draft (1979).

52. Wong, C. K. and Yao, A. C., "A combinatorial optimization problem related to data set allocation," *Rev. Francaise Automat. Informat. Recherche Operationelle Ser. Bleue* **10.5** (suppl.) (1976), 83-95.

53. Yao, A. C., "New algorithms for bin packing" *J. Assoc. Comput. Mach.* **27** (1980), 207-227.

54. Yue, P. C. and Wong, C. K., "On the optimality of the probability ranking scheme in storage applications," *J. Assoc. Comput. Mach.* **20** (1973), 624-633.

ADDITIONAL CONSTRAINTS IN THE GROUP THEORETICAL APPROACH TO INTEGER PROGRAMMING

V.FERRARI, S.GIULIANELLI, M.LUCERTINI

Istituto di Automatica e Centro di Studio dei Sistemi di Controllo e Calcolo Automatici del C.N.R. Via Eudossiana, 18 - 00184 Roma Italy

ABSTRACT

One of the most promising ways to obtain more efficient algorithms in integer programming is based on the determination of equivalent integer programming problems with a lower computational complexity.

In this paper we are concerned with equivalent problems obtained via a group theoretic approach and via the introduction of additional constraints. The procedure proposed consists in a manipulation of the ILP problem by adding a new unbinding constraint, in order to obtain a new problem and a new dual feasible basis such that the associated group problem has a computational complexity lower than the group problem associated to the original ILP problem.

1. INTRODUCTION

One of the most promising ways to obtain more efficient algorithms in integer programming is based on the determination of equivalent integer programming problems with a lower computational complexity.

By relaxing the nonegativity constraints on a set of basic variables in a linear programming problem (LP), an integer linear programming problem (ILP) can be reduced to a shortest route problem over a finite Abelian group (ILPC , integer linear programming over cones) [1,2,3].

Many recent works in integer programming have concentrated on the group theoretic approach and the related computational aspects [4,10,11,12,13]. This approach provides the solution of some asymptotic integer problems that either give directly the optimal solution of the ILP problem or can be utilized to obtain bounds for an enumeration algorithm built for non-asymptotic problems.

A simple way to reduce the computational complexity of the ILPC problem can be based on the observation that it is not necessary to choose the optimal LP basis in order to reduce the ILP to the ILPC problem. We can observe that there is a family of ILPC's corresponding to an ILP, one for each basis of the LP associated to the ILP. However only the dual feasible subfamily is useful; in fact if we start from a non dual feasible basis we obtain an unbounded ILPC. The complexity of ILPC's associated to different bases are in general different, then a suitable choice of the (dual feasible)basis can reduce the complexity of the ILPC problem. However, rigorous criteria of choice are difficult to point out and, on the other hand, there exists a greater probability that some entries of the LP basis vector, corresponding to the optimal solution of the ILPC problem, result in this case to be negative.

An interesting different approach approach consists in the generation of new constraints that can be added to the problem, that are redundant in a ILP sense, but are not so for the associated LP relaxation; with these additional constraints it is possible to explicitely derive the convex hull of a suitable subset of the set of all the constraints [3,9].

In this paper we are concerned with equivalent problems obtained via a group theoretic approach (see section 2) and via the introduction of additional constraints (see section 3). The procedure proposed consists of a manipulation of the ILP problem by adding a new unbinding constraint, in order to obtain a new problem and a new dual feasible basis such that the associated ILPC problem has a computational complexity lower than the ILPC problem associated to the original ILP problem.

The new ILPC problem is a relaxation of the original one and the basis variable vector corresponding to its optimal solution usually has some negative entries. However the branch and bound procedure utilized to find the optimal solution of the ILP results to be, under suitable hypotheses, more efficient than the procedure based on the classical group algorithms.

In section 4 a geometrical interpretation of the procedure is given. The main results, rigorously pointed out in section 5, are intuitively shown. Section 5 is the core of the paper, the theoretical results, concerning the reduction of the computational complexity via the introduction of a new constraint, are given and how to utilize them in a computational algorithm is also shown.

In many practical cases we are not interested in the "optimal" solution of the problem but in a "good" solution.

In section 6 a heuristic algorithm, based on the previous results, is pointed out. The solution obtained is feasible but, in general, not optimal. In section 7 some numerical results are given and some examples are presented to clarify the procedure.

2. FORMULATION OF THE GROUP PROBLEM

2.1. The general ILP problem can be written as:

$$(P') \qquad \max z(x') = c'x' \qquad x' \in F(P')$$

$$F(P') = \{x' : A'x' \leq b,\ x' \geq 0,\ x' \text{ integer}\}$$

with $A'(m \times n)$, $b(m \times 1)$, $c'(1 \times n)$, $x'(n \times 1)$. Let PR' be the problem obtained from P' dropping the integrity constraints:

$$(PR') \qquad \max z(x') = c'x' \qquad x' \in (PR')$$

$$F(PR') = \{x' : A'x' \leq b,\ x' \geq 0\}$$

Problem P' can be transformed, adding m slack variables, into the problem P

$$(P) \qquad \max z(x) = cx \qquad x \in F(P)$$

$$F(P) = \{x : Ax = b,\ x \geq 0,\ x \text{ integer}\}$$

with $A = [A' \mid I_m]$, $c = [c' \mid 0]$, $x = [x' \mid s]$.

In the same way a problem PR can be defined.

2.2. In order to obtain an ILPC problem corresponding to P we can relax P as follows (utilizing the same symbols to indicate the matrices and the corresponding sets of column indexes):

$$\min z'(x) = -(c_N - c_B B^{-1}N)x_N = d_N x_N$$

$$x_B + B^{-1}Nx_N = B^{-1}b$$

$$x_N \geq 0, \ (x_B, x_N) \text{ integer.}$$

with (B,N) a dual feasible basis of the problem PR and $A = [B \mid N]$ (in case after rearrangement of columns of the matrix A). From this problem it is possible to define the following group problem [15]:

$$\text{(PC)} \quad \min z'(x) = d_N x_N \qquad x_N \in F(PC)$$

$$F(PC) = \{x_N : B^{-1}Nx_N \equiv B^{-1}b \ (\text{mod } 1), \ x_N \geq 0, \ x_N \text{ integer}\}$$

As it is well known, [1,2], this group problem is a relaxation of P in which, for a given B, the nonnegativity restrictions on x_B are omitted. In x_N space the feasible solutions of PC, dropping the integrity constraints on x_N, correspond to the cone given by the nonegative orthant and if an ILPC has a feasible solution, it has an infinite number of feasible solutions. The convex hull of these solutions is, in this case, an unbounded polyhedron called the corner polyhedron.

There exists a one-to-one correspondence between variables and constraints. In particular each original variable x_j corresponds to the constraint $x_j \geq 0$ and each slack variable s_i corresponds to the ith constraint of the problem P'. Dropping the nonnegativity restrictions on a given variable corresponds to eliminating the associated constraint from the problem.

It is easy to see now that also in the x' space the feasible region of PC is a convex cone. In fact only the consrtaints associated to the nonbasic variables hold and all the corresponding hyperplanes pass through the vertex of F(PR') corresponding to (B,N).

2.3. An equivalent form of problem PC, more useful from some points of view, is the so called Smith canonical form [15]:

$$\min z'(x) = d_N x_N$$

$$\Delta \bar{x}_B + G x_N = g^o$$

$$x_N \geq 0 \quad (\bar{x}_B, x_N) \text{ integers}$$

where Δ is a diagonal matrix:

$$\Delta = RBC = \begin{bmatrix} \delta_1 & & & \\ & \delta_2 & & \\ & & \ddots & \\ & & & \delta_m \end{bmatrix}$$

with (R,C) integer unimodular square matrices, δ_i positive integers, δ_i divisor of $\delta_{i+1} (\forall i)$; $G = RN$; $g^o = Rb$; $\bar{x}_B = C^{-1} x_B$ (remark that [15] ($\bar{x}_B$ integer) $\Leftrightarrow$ (x_B integer)). The corresponding group problem in canonical form can be written as:

$$(PC_c) \quad \min z'(x_N) = d_N x_N \qquad x_N \in F(PC_c)$$

$$F(PC_c) = \{x_N : Gx_N \equiv g^o (\text{mod } \delta), x_N \geq 0, \ x_N \text{ integer}\}$$

where $\delta^T = \left[\delta_1 \mid \delta_2 \mid \ldots . \mid \delta_m\right]$.

It can be demonstrated that the problem PC and PC_c are completely equivalent [15], i.e. the associated Abelian groups are isomorphic; let φ be the group associated to PC_c.

2.4. The complexity of solution of a group problem depends on the order of the group φ, which can be shown to be equal to the determinant of B [15].

PROPOSITION 1. *The group problem* PC *(or* PC_c*) can be solved as a shortest path problem over a graph with* δ^* *nodes (elements of the finite Abelian group associated to* PC *or* PC_c*)*

$$\delta^* = |\det B| = \det \Delta = \prod_{i=1}^{m} \delta_i = (\text{order of } \varphi)$$

In this sense δ^* can be considered an index of the computational complexity of PC.

PROPOSITION 2. *If* (x_B^*, x_N^*) *is an optimal solution of* $PC(PC_c)$, $x_B^* = B^{-1}(b - Nx_N^*)$, $(x_B^* = C(g^o - Gx_N^*))$ *and* $x_B^* \geq 0$, *then* (x_B^*, x_N^*) *is an optimal solution of* P.

2.5. When some entries of x_B^* result to be negative, a branch and bound algorithm in which the subproblems are ILPC's can be used to solve the ILP [1].

Let PC^k be a subproblem generated in the branch and bound procedure. If PC^k is not fathomed it can be separated in (n+1) subproblems PC_j^k (j=0,1,...,n); the separation can be performed as follows:

$$F(PC^k) = \bigcup_{j=0}^{n} F(PC_j^k)$$

$$F(PC_o^k) = F(PC^k) \cap \{x_N | x_N = p^k\}$$

$$F(PC_j^k) = F(PC^k) \cap \{x_N | x_j \geq p_j^k + 1 \qquad j=1,\dots,n$$

where ($p_j^o = 0$, $\forall j$) and, denoting with $p_i^{k_j}$ the lower bounds of the problem PC_j^k,

$$p_j^{k_j} = p_j^k + 1 \;, \qquad p_i^{k_j} = p_i^k \;\; (i \neq j)$$

This separation is not a partition, however it is possible to avoid the solution of equivalent problems solving at each step only the subproblems PC_j^k $(j=0,1,\dots,j(k)-1)$ where:

$$j(k) = \begin{cases} \max\limits_{j \in R_k} & R_k \neq \emptyset \\ 1 & R_k = \emptyset \end{cases} \qquad R_k = \{j | p_j^k > 0\}$$

Remark that it is not necessary to solve an ILPC problem at each node of the branch and bound tree. In fact the problem PC^k differs from the original ILPC (PC^o) only for the right hand side vector of the problem.

Referring for instance to the canonical form, the problem PC^k is equal to the problem PC^o substituting g^o with

$$(g^o \oplus \sum_{j \in N} (-g^j) p_j^k)$$

where $\oplus$ denotes that sums are taken (mod δ) and g^j is the jth column of matrix G; obviously it is necessary to add p^k to the solution of PC^k obtained with the modified right hand side.

In practice we can calculate the shortest path tree related to the group φ (the same in all the subproblems!) at the beginning, and at each node we simply take the correspondent solution.

3. FORMULATION OF THE ILPC AUGMENTED PROBLEM

3.1. The new problem $\hat{P}$ (ILPCA problem) obtained from probelm P (expressed in Smith canonical form) by introducing an additional constraint can be written as follows:

$$(\hat{P}) \quad \min \hat{z}(x) = d_N x_N = d_{\hat{B}} \bar{x}_{\hat{B}} + d_{\hat{N}} x_{\hat{N}}; \qquad \bar{x} \in F(\hat{P})$$

$$F(\hat{P}) = \{\bar{x} : \hat{B}\bar{x}_{\hat{B}} + \hat{N}x_{\hat{N}} = \hat{b},\ x_{\hat{N}} \geq 0, (\bar{x}_{\hat{B}}, x_{\hat{N}}) \text{ integers}\}$$

where:

$$\hat{B} = \left[\begin{array}{c|c} \Delta & g^k \\ \hline t & h \end{array}\right]; \quad \hat{N} = \left[\begin{array}{c|c} G^k & 0 \\ \hline q & 1 \end{array}\right]; \quad \hat{b} = \left[\begin{array}{c} g^0 \\ \hline u \end{array}\right]$$

$$\bar{x} = \left[\begin{array}{c} \bar{x}_{\hat{B}} \\ \hline x_{\hat{N}} \end{array}\right]; \quad \bar{x}_{\hat{B}} = \left[\begin{array}{c} \bar{x}_B \\ \hline x_k \end{array}\right]; \quad x_{\hat{N}} = \left[\begin{array}{c} x_{Nk} \\ \hline x_{n+m+1} \end{array}\right]; \quad x_{\hat{B}} = \left[\begin{array}{c} x_B \\ \hline x_k \end{array}\right];$$

$$x_B = C\bar{x}_B \; ; \; x_{\hat{B}} = \hat{C}\bar{x}_{\hat{B}} \; ; \; \hat{C} = \left[\begin{array}{c|c} C & 0 \\ \hline 0 & 1 \end{array}\right];$$

$$d_{\hat{N}} = \left[\begin{array}{c|c} d_{Nk} & 0 \end{array}\right]; \; d_{\hat{B}} \left[\begin{array}{c|c} d_k & d_k \end{array}\right] = \left[\begin{array}{c|c} 0 & d_k \end{array}\right]$$

- x_k is an optimal nonbasic variable of problem P which we selected as a basic variable in problem $\hat{P}$;
- x_{Nk} is obtained from the vector x_N by deleting the x_k entry,

$$x_N^T = \left[x_{Nk}^T \,\vdots\, x_k \right] ;$$

- g^k is the column vector of G which corresponds to x_k ;
- G^k is obtained from the matrix G by deleting the g^k column,

$$G = \left[G^k \,\vdots\, g^k \right] ;$$

- x_{n+m+1} is the slack variable of the new constraint;
- t is an integer row vector $(1 \times m)$;
- q is an integer row vector $(1 \times (n-1))$;
- h and u are integers.

The group problem associated with $\hat{P}$ and with $(\hat{B},\hat{N})$ can be written as follows (for simplicity $\hat{PC}$ is not expressed in Smith canonical form):

$$(\hat{PC}) \qquad \min \hat{z}(x_{\hat{N}}) = \hat{d}_{\hat{N}} x_{\hat{N}}, \; x_{\hat{N}} \in F(\hat{PC})$$

$$F(\hat{PC}) = \left\{ x_{\hat{N}} : \hat{G} x_{\hat{N}} \equiv \hat{g}^{\circ} (\text{mod } 1), \; x_{\hat{N}} \geq 0, \; x_{\hat{N}} \text{ integer} \right\}$$

where $\hat{G} = \hat{B}^{-1}\hat{N}$, $\hat{g}^{\circ} = \hat{B}^{-1}\hat{b}$,

$$\hat{B}^{-1} = \frac{1}{\alpha} \begin{bmatrix} (\Delta^{-1}\alpha + \Delta^{-1} g^k t \Delta^{-1}) & -\Delta^{-1} g^k \\ -t\Delta^{-1} & 1 \end{bmatrix}$$

$$\alpha = \alpha(h,t,k) = h - t\Delta^{-1} g^k$$

$$\hat{d}_{\hat{N}} = d_{\hat{N}} - d_{\hat{B}}\hat{G}$$

The vector $x^*_{\hat{B}}(\hat{PC})$ corresponding to the optimal solution $x^*_{\hat{N}}(\hat{PC})$ of $\hat{PC}$ can now be calculated as:

$$x^*_{\hat{B}}(\hat{PC}) = \hat{C}(\hat{g}^o - \hat{G}x^*_{\hat{N}}(\hat{PC}))$$

$x^*_{\hat{B}}(\hat{PC})$ usually has some negative entries.

Following the same procedure as in 2.5, a branch and bound algorithm can be utilized to solve $\hat{P}$ (and then P).

As we have done for PC, we associate with $\hat{PC}$ a computational complexity index.

PROPOSITION 1'. *The group problem* $\hat{PC}$ *can be solved as a shortest path problem over a graph with* $\hat{\delta}^*$ *nodes:*

$$\hat{\delta}^* = |\det \hat{B}| = |(h - \sum_{i=1}^{m} \frac{g_i^k t_i}{\delta_i})|\delta^* = |(h - t\Delta^{-1}g^k)|\delta^* = |\alpha|\delta^*$$

REMARKS

1. The problem PC is not equivalent to $\hat{PC}$, the latter being a relaxation of the former because x_k belongs to N but does not belong to $\hat{N}$; consequently, $\hat{PC}$ x_k is not constrained to be nonnegative.
2. Given δ^* and k, $\hat{\delta}^*$ depends only on some coefficients of the new constraint (t and h).
3. The goal of reducing the computational complexity of problem $\hat{P}$ is achieved if we are able to construct the additional constraint so that $1 \leq \hat{\delta}^* < \delta^*$, i.e. if:

$$\frac{1}{\delta^*} \leq |\alpha| < 1$$

Note that from proposition 1', $\hat{\delta}^*$ must be nonnegative and integral; the constraint $\hat{\delta}^* \geq 1$ simply implies that the new basis matrix $\hat{B}$ is nonsingular.

4. Obviously, the new constraint must be such that some conditions are respected. These are a) each solution of P must be a solution of $\hat{P}$, i.e. $F(P) \subseteq F(\hat{P})$; b) moreover, as we want to utilize $\hat{B}$ as a basis for the solution of $\hat{P}$, it must be a dual feasible one. In sections 3.3, 3.2. the consequences of introducing conditions a) and b) will be discussed.

5. x_{n+m+1} cannot be chosen to enter the basis, as well as any other x_j such that $g^j = 0$, since it would follow $|\alpha| = |h| \geq 1$.

THEOREM 1. $(\exists |\alpha| = \gamma/\delta_m, \gamma$ *integer*, $1 \leq \gamma \leq \delta_m) \Leftrightarrow (\exists k : GCD(\pi_k) | \gamma)$ *with* $GCD(\pi_k) \triangleq GCD\{\delta_m, -\delta'_1 g_1^k, -\delta'_2 g_2^k, \ldots, -\delta'_m g_m^k\}$; $(\delta'_i = \frac{\delta_m}{\delta_i}$, integer $\forall i)$ (i.e. *it is possible, with suitable values of* h *and* t, *to obtain a value of* α *equal to* γ/δ_m *if and only if there exists a* k *such the greatest common divisor of the given elements divides* γ).

PROOF. This result is a strainghtforward consequence of a well known number theory theorem [17], stating that the least positive value of a linear form $\sum_j b_j y_j$, where $\{b_j\}$ are given integers and $\{y_j\}$ range over all integers, is the GCD of $\{b_j\}$, and of the definition of Δ and α

$$\alpha(h,t,k) = \frac{h\delta_m - \sum_{i=1}^{m} t_i \delta'_i g_i^k}{\delta_m} \quad ◀$$

COROLLARY 1. $|\alpha(h,t,k)| \geq \frac{1}{\delta_m}$ $\quad (\hat{\delta}^* \geq \delta^*/\delta_m)$

PROOF. It is a direct consequence of Theorem 1. From

the definition of α, it follows that, as α cannot be zero, the minimum value of $|\alpha|$ is $1/\delta_m$. ◀

As a consequence it is possible to reduce the determinant by $1/\delta_m$ only if theorem 2 holds with $\gamma = 1$.

COROLLARY 2.

$$\alpha^*(k) \triangleq \min_{(h,t)\,\text{integers}} |\alpha(h,t,k)| = \mathrm{GCD}(\pi_k)/\delta_m$$

$$\alpha^*(k^*) \triangleq \min_{k \in N} \alpha^*(k) = \min_{k \in N} \mathrm{GCD}(\pi_k)/\delta_m$$

COROLLARY 3. $(\hat{\delta}^* = 1) \Rightarrow$ (*φ is cyclic*) (i.e. *φ cyclic is a necessary condition to obtain* $\hat{\delta}^* = 1$)

PROOF. If (and only if) φ is cyclic $\delta_m = \delta^*$ [15], which is a necessary condition for $\hat{\delta}^* = 1$.◀

COROLLARY 4. *For each* (k,i) *such that* $|g_i^k| < \delta_i$ *it is always possible to reduce* δ^* *at least to*

$$\hat{\delta}^* = \delta^* \frac{|g_i^k|}{\delta_i}$$

COROLLARY 5. *If* $\delta_i' > 1$ *and* $\delta_j' = 1$ $(j=i+1,\dots,m)$:

$$\left(\mathrm{GCD}\left\{\delta_i', g_{i+1}^k, g_{i+2}^k, \dots, g_m^k\right\} > 1\right) \Rightarrow \left(\alpha > \frac{1}{\delta_m}\right)$$

The proof of corollaries 4 and 5 are omitted in this work because they require a relatively large space but they are conceptually simple and a direct consequence of Theorem 1.

3.2. Dual feasibility conditions for a basic solution in LP are well known [1]; for the problem $\hat{PR}$ (obtained from $\hat{P}$ dropping the integrity constraints) and $(\hat{B},\hat{N})$ they are:

$$\hat{d}_{\hat{N}} = d_{\hat{N}} - d_{\hat{B}}\hat{G} \geq 0$$

Let g^j be the column of G^k corresponding to the variable x_j $(j \in N^k)$, and let $\langle e \rangle$ be the smallest integer greater than or equal to e, [e] the greatest integer less than or equal to e.

THEOREM 2. *Given* (h,t,k), $\hat{d}_{\hat{N}} \geq 0$ *if:*

$$\alpha(h,t,k) < 0$$

$$g_j \geq \langle \bar{q}_j^k \rangle$$

$$\bar{q}_j^k = (t\Delta^{-1}g^j + \frac{d_j}{d_k}\alpha(h,t,k)), \quad j \in N^k$$

PROOF. The dual feasibility condition $\hat{d}_{\hat{N}} \geq 0$ can be written:

$$\hat{d}_{n+m+1} = -\frac{d_k}{\alpha(h,t,k)} \geq 0$$

$$\hat{d}_j = d_j - \frac{q_j - t\Delta^{-1}g^j}{\alpha(h,t,k)} d_k \geq 0, \quad j \in N^k$$

From $d \geq 0$ and q integer, supposing $d_k > 0$ (if $d_k = 0$ the previous relations are trivially satisfied), we can infer the thesis. ◀

As a conclusion we can assert that it is always possible to select a sufficiently large value of q, such that the dual feasibility constraints are satisfied.

3.3. Moreover, given t,h and q, it is always possible to choose u such that $F(\hat{P}) \supseteq F(P)$. More precisely let w^* be the optimal value of the objective function of the following pro blem.

(Q) $\max\ w(\bar{x}) = t\bar{x}_B + hx_k + qx_{Nk}$, $\bar{x}$ such that $x \in F(P)$

the following proposition holds.

PROPOSITION 3.

$$(u \geq w^*) \Leftrightarrow F(P) \subseteq F(\hat{P}))$$

PROOF. Proposition 3 is obvious if we note that the new constraint can be written as $w(\bar{x}) \leq u$.◂

Let now w_R^* be the optimal solution of QR.

(QR) $\max\ w(\bar{x})$, $\bar{x}$ such that $x \in F(PR)$

From Proposition 3 it follows:

PROPOSITION 3'. *A sufficient condition to assure* $F(\hat{P}) \supseteq F(P)$ *is that* $u \geq [w_R^*]$ $(\geq w^*)$; *in addition* $(u \geq w_R^*) \Leftrightarrow$ $(F(PR) \subseteq F(P\hat{R}))$.

Let $u^*(t,h,q)$ indicate the value $|w_R^*|$.

REMARK. If F(P) is an unbounded region, problem Q may have no finite solution. With a different choice of the variable x_k (entering the basis) problem Q may possibly be solved, anyway in section 5 a different way to face this problem will be shown.

4. THE GEOMETRY OF THE PROBLEM

A geometrical interpretation of the approach proposed in this work is particularly important in order to fully understand the theoretical results given in section 5 and their practical utilization in a improved group algorithm.

As we have said before, the ILPCA is a relaxation of the ILPC. In fact the variable x_k introduced in the basis corresponds to a face of the cone, and relaxing the non-negativity constraint implies dropping this face; adding the new constraint a new cone is formed, containing all the feasible solutions of P and, usually, new non-feasible solutions. It is reasonable to think that the "closer" the new constraint V_a is to the dropped one V_d , the smaller the number of new nonfeasible solutions introduced in the problem becomes. A good index to measure the "closeness" between the two constraints seems to be the angle $\beta(V_d,V_a)$ between the normals to the two hyperplanes V_d and V_a (briefly "angle between the constraints"):

$$\text{(A)} \qquad \cos \beta(V_d,V_a) = \frac{[a^d, a^s]}{|a^d|\cdot|a^a|}$$

where a^d and a^a are the coefficient vectors of the constraints V_d and V_a ; $[\cdot,\cdot]$ indicates the inner product between two vectors.

In the following we show in a simple two dimensional example (fig.1) how the new cone varies as V_a varies. We are particularly concerned with the role of the coefficients q_j ; (t,h) is assigned with regard to the value $\hat{\delta}^*$ and u can be fixed following section (3.3). All the conclusions can be easily extended to the n-dimensional case.

DEFINITION. *A constraint is said to be in "geometrical form" if all the slack variables except its own have zero coefficients.*

Let now $x_N^T = \left[x_v \mid x_w \right]$ be a 2-vector; let x_v correspond to the constraint V_v (BC), x_w to V_w (BA). If x_v enters the basis, the coefficients (t,h) and q_w of the new constraint V_a must

be assigned; we assign t and h according to Theorem 1, then, disregarding for a moment the dual feasibility condition, fix q_w at a large negative value $-\varepsilon$.
If x_w is a slack variable, we can restore the geometrical form by adding ε times V_m to V_a. Then it can be easily seen that $\beta(V_w,V_a)$ is very small (see also formula A). If we increase q_w, V_a rotates; if q_w has a large positive value (ε), $\beta(V_w,V_a) \simeq 180°$.

We will prove soon that the rotation is as shown in the figure. When varying q_w from $-\varepsilon$ to $+\varepsilon$, V_a rotates "in jerks" round the vertex B (optimal solution of PR), and $\beta(V_v,V_a)$ decreases. At a certain moment the centre of rotation moves to C, $\beta(V_v, V_a)$ increases again, and the vertex of the cone $F(\hat{PC})$ goes farther and farther from B, remaining on the prolongation of BA.

If x_w is a x' variable (that is, AB is a coordinate axis), everything is quite similar: if $q_w = -\varepsilon << 0$, V_a forms a small angle with $-x_w \leq 0$, which is just V_w, etc.

The condition of dual feasibility has now a simple interpretation. If, for instance, the objective function has a slope between BC" and BC"' (fig. 1b), the least feasible value for q_w is the one corresponding to BC"'; for a smaller value, the objective function would always intersect $F(\hat{PC})$, i.e. $\hat{PC}$ would not have a finite optimum solution. If the z(x) has the same slope as AB (this happens if $d_v = 0$), any value of q_w is feasible. If the z(x) lies between BC"" and BC, the least feasible value for q_w corresponds to B'C (fig. 1c). Now it can be easily proved that the rotation of V_a (related to the increase of q_w) occurs in the shown direction: on the contrary, the z(x) would always intersect the cone, i.e. dual feasibility would be impossible.

With regard to the extent of the relaxation, it is obvious that the best choice for Va is BC if admissible, other wise B'C. But it must be noted that the constraints passing through the vertex B of the old cone are in general more efficient than the constraints passing through other vertices of the polyhedron.

It may happen that no choice of q_w leads to a $\beta(V_v, V_a)$ that is small enough. One might suppose that a different assignment of t and h deserves then to be tested, but it will be proved in the next section that this is not the case.

The set of constraints V_a can be "thickened" through a different observation. A fractional value of q_w would obviously allow a better approximation of V_v; this can actually be made, provided that all the coefficients of V_a are then multiplied by the denominator q_D of q_w. A disadvantage of this technique is that the determinant $\hat{\delta}^*$ increases of the same amount ($\bar{\delta} = q_D \cdot \hat{\delta}^*$). It is interesting to note that we can obtain the eliminated constraint only by increasing $\hat{\delta}^*$ to the original value δ^*, i.e. without any computational advantage.

5. THEORETICAL RESULTS

5.1. From section 3 it follows that, given k (index of the variable entering the basis), (t,h) must be a solution of the linear diophantine equation

$$\text{(D)} \qquad \delta_m h - \sum_{i=1}^{m} \delta_i' g_i^k t_i = \frac{\delta_m}{\delta^*} \bar{\delta} \quad (=\gamma)$$

with $\bar{\delta}$ the value of the determinant of $\hat{B}$ that we want to obtain (in particular this holds with the right hand side equal to $\delta_m \alpha^*(k)$).

If this equation has one solution (t_c, h_c), then it has an infinite number of solutions all expressed by the formula

[17]:

$$\left[t \;\vdots\; h \right] = \left[t_o \;\vdots\; h_o \right] + vW$$

with W(m,m+1) a suitable integer matrix and v an arbitrary integer m-vector.

Let $\aleph^k(\bar{\delta})$ and $\mathcal{H}^k(t,h)$ be the following sets:

$\aleph^k(\bar{\delta})$ = {(t,h): D holds}

$\mathcal{H}^k(t,h)$ = {set of the additional constraints $V_a(t,h,q)$ in geometrical form (i.e. in the x' space) such that the dual feasibility conditions hold ($q \geq \bar{q}^k$), and q is an integer vector}.

THEOREM 3. $\mathcal{H}^k(t,h)$ *is an invariant for* $(t,h) \in \aleph^k(\bar{\delta})$.

PROOF. The proof, rather long but not difficult, will be given in two parts.

1) Fixed x_k entering the basis such that $d_k \neq 0$, we choose a aolution $(t \,\vdots\, h) \in \aleph^k(\bar{\delta})$ of the diophantine equation, and assign:

$$q_j = \langle \bar{q}_j^k \rangle = \langle t\Delta^{-1}g^j + \alpha(h,t,k)d_j/d_k \rangle$$

(the least value satisfying the dual feasibility conditions (section 3.2)). Let us prove now that the geometrical form of the additional constraint V_a does not depend on the particular choice of $(t \,\vdots\, h)$, i.e. on the arbitrary vector v. The proof is nothing but the synthesis of V_a.

The matrix $A = (A' \,\vdots\, I_m)$ of the problem P is rearranged as follows (p is the number of original variables x' in the basis B):

$$(1)\qquad (B \vdots N) = \left(\begin{array}{c|c|c|c} B_{cx'} & B_{cs} & N_{cx'} & N_{cs} \\ \hline B_{rx'} & B_{rs} & N_{rx'} & N_{rs} \end{array}\right) \qquad \left(\begin{array}{c|c|c|c} B_{cx'} & 0 & N_{cx'} & I_p \\ \hline B_{rx'} & I_{m-p} & N_{rx'} & 0 \end{array}\right)$$

where the indexes x' and s indicate respectively original and slack variables; c and r indicate "cone" and "relaxed" constraints (the upper constraints belong to the cone F(PC) because their slack variables are nonbasic, the lower constraints do not because their slack variables are basic). The vectors b and c, of course, are rearranged in the same way.

We calculate now Δ = RBC and G = RN, then we add V_a. $(t \vdots h)$ is a solution of (D), i.e.

$$(2)\qquad \begin{aligned} t &= t_o + vW^{m+1} \\ h &= t\Delta^{-1}g^k - \gamma/\delta_m \end{aligned}$$

where W^{m+1} is obtained from W by dropping its column w^{m+1}. Then V_a is completely defined by t,h and $q = \langle \bar{q}^k \rangle$ (the coefficient u being defined without ambiguity). However from a computational point of view it is useful to write $[h \vdots q]$ as a unique vector $\tilde{q}$

$$\tilde{q} = \langle t\Delta^{-1}G - d_N\gamma/d_k\delta_m \rangle = [q_1 \ldots q_{k-1} h q_{k+1} \ldots]$$

However it is necessary to give another expression of q. Recall

$$\delta_m h - \delta_m t\Delta^{-1}g^k = -\gamma$$

and its general solution

$$\delta_m h_o + \delta_m v w^{m+1} - \delta_m t_o \Delta^{-1} g^k - \delta_m v w^{m+1} \Delta^{-1} g^k = -\gamma$$

The sum of the first and third term on the left makes $-\gamma$, hence

$$v w^{m+1} \Delta^{-1} g^k = v w^{m+1}$$

and $v w^{m+1} \Delta^{-1} g^k$ is integer. Then

(3) $$\tilde{q} = \langle t\Delta^{-1} G - d_N \gamma / d_k \delta_m \rangle = \langle t_o \Delta^{-1} G - d_N \gamma / d_k \delta_m \rangle + v w^{m+1} \Delta^{-1} G$$

The additional constraint is $(t' = tC^{-1})$

$$t' x_B + \tilde{q} x_N \leq u^*$$

$$(t'_{x'} \,\vdots\, \tilde{q}_{x'}) x' + (\tilde{q}_s \,\vdots\, t'_s) s \leq u^*$$

It can be reduced to its geometrical form by suitably adding the other original constraints to it, obtaining

(4) $$\left[(t'_{x'} \,\vdots\, \tilde{q}_{x'}) - (\tilde{q}_s \,\vdots\, t'_s) \begin{pmatrix} B_{cx'} & N_{cx'} \\ \hline B_{rx'} & N_{rx'} \end{pmatrix} \right] x' \leq u^* - (\tilde{q}_s \vdots t'_s) b = u'$$

By developing this expression by (2),(3) and the partition (1), it can be shown that all the terms in which the arbitrary vector v appears annul each other. The first part of the proof is then completed.

2) Now let us fix an integer n-vector f, such that

$$f_j \geq 0 \qquad j \neq k$$
$$f_k = 0$$

If the vector $(h \vdots q)$ is assigned as

$$(h \vdots q) = \tilde{q} + f$$

both the dual feasibility conditions and the reducibility condition hold, and the previous discussion holds again too: the geometrical form of V_a depends of course on f, but not on $(t \vdots h)$. Hence the theorem is proved. If $d_k = 0$, this only means that any integer value of q is admissible.◀

5.2. Theorem 3 proves the statement, made in section 4, that any choice of (t,h) satisfying the equation (D) is equivalent with respect to the minimization of $\beta(V_d,V_a)$: the angle β depends only on q. More exactly, varying the choice of (t,h) (provided that $\hat{\delta}^*$ does not change) implies only a shift in the mapping from the set of vectors q (Θ) to the set of the constraints V_a ($\mathcal{H}$).
Remark that there are no more reasons to consider only integer values of q: as we have said at the end of section 4, fractional values of q can be extremely useful for "thickening" the set $\mathcal{H}$. In this aim we can define Θ and redefine $\mathcal{H}$ as follows:

$\Theta^k(t,h)$ = {set of the rational vectors q such that $q \geq \bar{q}^k$ and $q_D \cdot \hat{\delta}^* \leq \delta^*$ ($q_D \cdot |\alpha| \leq 1$) with q_D equal to the least common denominator of $\{q_i\}$; (t,h) integers}.

$\mathcal{H}^k(t,h)$ = {set of the additional constraints $V_a(t,h,q)$ in geometrical form such that $q \in \Theta^k(t,h)$}.

In the following theorems we examine the relation between the sets Θ and $\mathcal{H}$ more deeply.

THEOREM 4. *The map from $\Theta^k(t,h)$ to $\mathcal{H}^k(t,h)$ is surjective and injective.*

PROOF. The first statement is obvious. To show that the map is injective, we consider again the expression (4) of theorem 3. Let q' and q" be two values of $\tilde{q}$ mapping to the same constraint V_a, then

$$\left[t'_{x'} \,\middle|\, q'_{x'}\right] - \left[q'_s \,\middle|\, t'_s\right] \left[\begin{array}{c|c} B_{cx'} & N_{cx'} \\ \hline B_{rx'} & N_{rx'} \end{array}\right] = \left[t'_{x'} \,\middle|\, q''_{x'}\right] - \left[q''_s \,\middle|\, t'_s\right] \left[\begin{array}{c|c} B_{cx'} & N_{cx'} \\ \hline B_{rx'} & N_{rx'} \end{array}\right]$$

$$\left[0 \,\middle|\, q'_{x'} - q''_{x'}\right] = \left[q'_s - q''_s \,\middle|\, 0\right] \left[\begin{array}{c|c} B_{cx'} & N_{cx'} \\ \hline B_{rx'} & N_{rx'} \end{array}\right]$$

Hence:

$$0 = \left[\;'_s - \;''_s\right] B_{cx'}$$

$$\left[\;'_{x'} - \;''_{x'}\right] = \left[\;'_s - \;''_s\right] N_{cx'}$$

and, remembering that $|\det B_{cx'}| = \delta^*$, $q' = q''$ ◂

THEOREM 5. *Given* k *and* $(t,h) \in \aleph^k(\bar{\delta})$, *let us choose*

$$q = \bar{q}^k = t\Delta^{-1}G^k + \alpha(h,t,k)\,\frac{d_{Nk}}{d_k}$$

Then V_a *is parallel to the objective function.*

PROOF. The proof, long but quite simple, is omitted. It only consists in substituting $\bar{q}^k$ inside the expression (4) of V_a, and developing the calculation (the coefficients of V_a result to be equal to $\frac{\alpha}{d_k}$ c). Remark that $\bar{q}^k$ is usually non integer. ◂

THEOREM 6. *Given* k *and* $(t,h) \in \aleph^k(\bar{\delta})$, *let us choose*

$$q = \bar{\bar{q}}^k = t\Delta^{-1}G^k$$

Then V_a *is parallel to the dropped constraint* V_d, i.e. $\beta(V_d,V_a) = 0$.

PROOF. The same as theorem 5. Remark that in this case $\bar{\bar{q}}^k_D\hat{\delta}^* = \delta^*$ with $\bar{\bar{q}}^k_D$ least common denominator of $\{\bar{\bar{q}}^k_i\}$.◂

COROLLARY 6. *The function of* q $\beta(V_a(\bar{t},\bar{h},q),V_d)$ *has a unique minimum (in general non integer)* $\beta = 0$ *in* $q = \bar{\bar{q}}^k$.

PROOF. Obvious considering theorems 4 and 6.◂

COROLLARY 7. $\bar{q}^k \leq \bar{\bar{q}}^k$.

PROOF. In fact $\bar{\bar{q}}^k - \bar{q}^k = -\alpha d_Nk/d_k \geq 0$ (see theorem 3). The equality occurs for those elements of d_Nk equal to zero, i.e. only if the solution is degenerate. ◂

For a simple formulation of the next theorem, we assign now $u = w^*$ rather than $u = [w^*]$ (see section 3.3):

THEOREM 7. *A dual feasible additional unbinding constraint* $V_a(t,h,q) = u$ *(with* $u = w^*$*) passes through the vertex of the original cone if and only if*

$$\bar{q}^k \leq q \leq \bar{\bar{q}}^k.$$

PROOF. $(\bar{q}^k \leq q)$ is only the dual feasibility condition, let us observe now that

$$\bar{\bar{q}}^k = t\Delta^{-1}G^k = tC^{-1}B^{-1}R^{-1}RN^k = t'B^{-1}N^k$$

and consider the condition

$$t'B^{-1}(N^k|n^k) - (q|h) \geq 0$$

This is nothing but the condition of dual feasibility of B, optimal basis of PR, when the objective function is $z = [t' \vdots q \vdots h]x' = V_a$, i.e. the necessary and sufficient condition for which $V_a = z^*$ passes through x'^*(PR), vertex of the original cone. It can be split into the two conditions

$$t'B^{-1}n^k - h \geq 0$$

$$t'B^{-1}N^k - q \geq 0$$

where the first is always satisfied because

$$h \triangleq t\Delta^{-1}g^k - \alpha < t\Delta^{-1}g^k = t'B^{-1}n^k$$

and the second is the hypothesis of the theorem. ◂

Remark that the difference between $\bar{\bar{q}}^k$ and $\bar{q}^k$ is $-\alpha d_N^k/d_k$. Then as $|\alpha|$ decreases, it becomes unlikely to find an integer vector in the interval $[\bar{q}^k, \bar{\bar{q}}^k]$. In this case we can either utilize a fractional value of q increasing the value of the determinant, or utilize a constraint not passing through the vertex, or both. The second procedure is, in general, less efficient (but in many cases the only possible), because of the large number of new integer solutions introduced into the problem, even for small values of the angle $\beta(V_a, V_d)$.
Remark also that, if $d_k = 0$ and $d_j \neq 0 \quad \forall j \neq k$, then $\bar{q}^k = -\infty$. Then it is very simple to find an integer vector q such that V_a passes through the vertex, for instance $q = [\bar{\bar{q}}^k]$ (also if this choice does not minimize, in general, β); the algorithm is in this case extremely efficient.

6. A HEURISTIC ALGORITHM

In many practical cases we are interested in a "good" solution of the problem rather than in the optimal solution. The procedure proposed in this work is particularly suitable for this purpose.

DEFINITION. *A problem* P' *is called a restriction of a problem* P" *if* $F(P') \subset F(P")$ *and the objective function is the same.*

Let η be the set of the vectors $\{V_i\}$ defining the original cone (constraints corresponding to the nonbasic variables) and η' be the set $\eta' = (V_a \cup (\eta - V_d))$ defining the new cone (remark that both η and η' are bases of $\mathbf{R}^n$).
Following the definitions of the previous section with $V_a(t,h,q,u) = V_a(q)$ defined by $t = t(k,\gamma), h = h(k,\gamma)$, $u = u^V(q)$ (value of u obtained with the previous values of (t,h) imposing that V_a pass through the vertex of the cone) we can point out the following result:

THEOREM 8. *The cone defined by* η' *is a restriction of the cone defined by* η *if and only if* $q \geq \bar{q}$.

PROOF.. It is an obvious consequence of Theorem 7. ◄

Remark that the part of the cone η eliminated (i.e. not belonging to η') can contain all the integer feasible solutions of the problem (i.e. all the solutions such that the basic variables are non-negative).

7. A NUMERICAL PROCEDURE

7.1. Following the results presented in the previous sections, a solution algorithm can be organized in many different ways. The main problem is in general the dimension of the branch and bound tree spanned because of the negative x_B solution of the group problem; in fact in many cases x_B obtained as solution of the ILPCA has large negative entries and the branch and bound procedure utilized in order to restore x_B non-negative is very heavy from a computational point of view. In this section an outline of a solution algorithm is presented, the algorithm is organized as follows:

- given a group problem to solve as an ILP problem via a branch and bound procedure (section 2.5, let P be this problem) we impose the additional condition of continuing the solution procedure only if all the nodes selected are such that (let h_j be the node selected and h its father):

  $$(\rho^h(P) - \rho^{h_j}(P))/\rho^h(P) \geq \eta$$

  with $\rho^h(P)$ absolute value of the sum of the negative entries of x_B in the node h of the branch and bound tree spanned solving P and η equal to the minimal accepted reduction of $\rho^h(P)$; let (solve P) indicate this procedure;
- the maximum order of the group that we accept in (solve P) is fixed (let δ_R^* be this value); in each step of the algorithm "thickening procedures" are utilized (see the end of section 4) in order to find better constraints, i.e. constraints V_a such that $\beta(V_a, V_d)$ will be smaller; in other words we accept to increase $\hat{\delta}$ at most to δ_R^* if we decrease β (obviously $\delta_R^* < \delta^*$);
- a relaxation of the cone such that the new cone has the same vertex of the old one is considered first (if it exists, i.e. if there exist constraints satisfying $\bar{q} \leq q \leq \bar{\bar{q}}$ and if such constraints form an angle $\beta(V_a, V_d)$ small enough, $\beta \leq \Theta_R^V$ with Θ_R^V maximum angle accepted), it can be intuitively shown (see section 4) that this is in general convenient with respect to a constraint V_a not passing

through the vertex;

- if the previous step fails to solve the problem, we take into account constraints V_a not passing through the original vertex, in this case we obviously have stronger constraints on β ($\beta \leq \Theta_R \leq \Theta_R^V$);
- if the two previous steps fail to solve the problem, we try to obtain an approximate solution, in practice we solve the same problem as in the previous step but imposing that V_a pass through the original vertex (i.e. with a different value of u); with these constraints a section of the original cone, containing possibly the optimal solution, is eliminated, but the new problem is considerably easier to solve; in particular if $q \geq \bar{\bar{q}}$ no new integer solutions are introduced in the problem (see section 6), let $\bar{\Theta}_R^V$ be the value of the maximum angle accepted in this case.

In order to simplify the exposition, the algorithm outlined in the following is not "optimized" from a computational standpoint. Some computations are not strictly necessary and some parameters that in general can be different, a different steps of the algorithm are assumed to be equal. Let be given:

- a solution of $D(t(k,\gamma)$ and $h(k,\gamma))$
- $K = \{k : \alpha^*(k) \leq \delta_R^*/\delta^*,\ k \in N\}$
- $I_k = \{i : i\alpha^*(k) \leq \delta_R^*/\delta^*,\ i \geq 1,\ i \text{ integer}\},\ k \in K$
- $\beta' \triangleq \beta(k,\hat{\hat{q}},Q^V,i^*)$ minimum angle between V_d and V_a for $q \in Q^V(k,i) \triangleq \{q : \bar{q}^k \leq q \leq \bar{\bar{q}}^k,\ q = q'/i,\ q' \text{ integer}\}$ and $i \in I_k$, obtained via a method of local variations [18], with $\hat{q}$ as initial value, step lenght in each iteration equal to $(1,1/2,\ldots,1/i^*(k))$ $(i^*(k)=\max\{i : i \in I_k\})$;
  $\beta = \beta(k,\hat{q},Q,i^*)$ the same as β' for $q \in Q(k,i) = \{q : q^{-k} \leq q \leq q^M,\ q = q'/i,\ q' \text{ integer}\}$, q^M a sufficiently large

value and $i \in I_k$, $\hat{q}$ as initial value; let (q_β, q'_β) be the values of q obtained in the previous minimizations, $(\gamma_\beta, \gamma'_\beta)$ the corresponding values of γ.

ALGORITHM

Step 0. Given: Δ, G, g^o, d

Step 1. Compute: $\alpha^*(k)(k \in N)$; $K; i^*(k), (\bar{q}^k, \bar{\bar{q}}^k), \hat{q}^k(k \in K)$; $Q(k,i), Q^V(k,i)(i \in I_k, k \in K)$.
If $\bigcup_{k \in K} \bigcup_{i \in I_k} Q^V(k,i) \neq \emptyset$ go to 2; otherwise go to 4

Step 2. Compute $(\beta', q'_\beta)(\forall k : \bigcup_{i \in I_k} Q^V(k,i) \neq 0)$ and memorize the results in a list L^V (ordered for increasing values of β') only if $\beta' \leq \Theta_R^V$. If L^V is empty go to 4; otherwise go to 3.

Step 3. Following the list L^V (i.e. beginning from the smaller values of the angle β') call (solve P) with $u = u^V(q'_\beta)$. For the first k such that the problem is solved memorize the solution and go to 7; otherwise go to 4.

Step 4. Compute (β, q_β) $(\forall k \in K)$ and memorize the results in two lists (ordered for increasing values of β) L^a if $\beta' \leq \Theta_R$, L^b if $\beta' \leq \bar{\Theta}_R$. If L^a is empty go to 6; otherwise go to 5.

Step 5. Following the list L^a call (solve P) with $u = u^*(t(k, \gamma_\beta), h(k,\gamma_\beta), q_\beta)$. For the first k such that the problem is solved memorize the solution and go to 7; otherwise go to 6.

Step 6. (Heuristic procedure). Following the list L^b call (solve P) with $u = u^V(q_\beta)$. For the first k such that the problem is solved memorize the approximate solution and go to 7; otherwise go to 7 (without a solution memorized).

Step 7. Terminate. If there exists a solution memorized the problem is solved (completely or approximately).

7.2. In this section a numerical example is solved, in order to show the main lines of the procedure. Let the problem P be

$$\max z = 3x_1 + x_2$$

$$(P)\quad \begin{cases} 4x_1 - 2x_2 + x_3 = 7 & (V_3) \\ 5x_1 + 14x_2 + x_4 = 102 & (V_4) \end{cases}$$

$$x_1, x_2 \geq 0 \ (V_1, V_2); \ x_1, x_2 \text{ integer}$$

It follows:

$$B = \begin{pmatrix} 4 & -2 \\ 5 & 14 \end{pmatrix} \quad N = \begin{pmatrix} 1 & 0 \\ 0 & 1 \end{pmatrix} \quad \delta^* = 66$$

$$x^*(PR) = \left(\frac{302}{66}, \frac{373}{66}\right), \quad d_N = \left(\frac{37}{66}, \frac{10}{66}\right)$$

$$R = \begin{pmatrix} -1 & 1 \\ 5 & -4 \end{pmatrix} \quad C = \begin{pmatrix} 1 & 16 \\ 0 & -1 \end{pmatrix} = C^{-1} \qquad \Delta = \begin{pmatrix} 1 & 0 \\ 0 & 66 \end{pmatrix}$$

$$RN = G = \begin{pmatrix} -1 & 1 \\ 5 & -4 \end{pmatrix} \qquad Rb = g^0 = \begin{pmatrix} 95 \\ -373 \end{pmatrix}$$

$$\pi_3 = \{66,\ 66,\ -5\} \qquad GCD(\pi_3) = 1 \quad (\hat{\delta}^* \geq 1)$$

$$\pi_4 = \{66,\ -66,\ 4\} \qquad GCD(\pi_4) = 2 \quad (\hat{\delta}^* \geq 2)$$

If x_3 enters the basis, we can assign

$$t = (-1 \quad -13) \qquad h = 0 \;\Rightarrow\; \alpha = -\frac{1}{66}, \quad \hat{\delta}^* = 1$$

$$\bar{q}^3 = \bar{q}_4 = -\frac{528}{2442} \simeq -0{,}216 \qquad \bar{\bar{q}}^3 = \bar{\bar{q}}_4 = -\frac{518}{2442} \simeq -0{,}212$$

Unfortunately, no fractional value of q_4 , exists in the interval $[\bar{q}_4\ \bar{\bar{q}}_4]$; such that $\hat{\delta}^* \leq \delta^*_R$, if we choose $q_4 = -\frac{1}{5}$ it results $(t' = tC^{-1} = (-1 \quad -3))$:

$$V_a = -x_1 - 3x_2 - \frac{1}{5}x_4 + x_5 = u$$

$$V_a = -\frac{1}{5}x_2 + x_5 = u + \ldots \text{ (in geometrical form)}$$

and $\beta(V_d,V_a) = \beta(V_3,V_a) \simeq 63°$, too big.

If x_4 enters basis:

$$t = (1 \quad 16) \qquad h = 0 \qquad t' = (1 \quad 0)$$

$$\bar{q}^4 = \bar{q}_3 = 1/10$$

$$\bar{\bar{q}}^4 = \bar{\bar{q}}_3 = \frac{14}{66} \simeq 0{,}212$$

	x_1	x_2	x_4	x_3	x_5
95	1	0	1	-1	0
-373	0	66	-4	5	0
u	t_1	t_2	h	g_3	1

In the next table we show the coefficients of V_a (in geometrical form) for different values of q_3; also $\beta(V_4,V_a)$ (in degrees) and $q_D\hat{\delta}^*$ are given.

q_3	V_a	β	$q_D\hat{\delta}^*$
$0^{(*)}$	$x_1 \qquad x_5 = \ldots$	70,3	2
$(\bar{q}_3 =)\frac{1}{10}$	$\frac{6}{10}x_1 + \frac{2}{10}x_2 + x_5 = \ldots$	51,9	20
$\frac{1}{5}$	$\frac{2}{10}x_1 + \frac{4}{10}x_2 + x_5 = \ldots$	6,9	10
$(\bar{\bar{q}}_3 =)\frac{7}{33}$	$\frac{10}{66}x_1 + \frac{28}{66}x_2 + x_5 = \ldots$	0	66
1/4	$\frac{1}{2}x_2 + x_5 = \ldots$	19,6	8
1/3	$-\frac{1}{3}x_1 + \frac{2}{3}x_2 + x_5 = \ldots$	46,2	6
1	$-3x_1 + 2x_2 + x_5 = \ldots$	75,9	2

(*) not feasible.

We solve the example for two different choices of q_3, $q_3=1/5$ and $q_3 = 1/4$. With the former V_a passes through the vertex, with the latter $q_D\hat{\delta}^*$ is a bit smaller (8 rather than 10).

1) $q_3 = \frac{1}{5}$

Then, multiplying all the coefficients of V_a (except the one of x_5) by 5,

$t = (5 \quad 80) \qquad h = 0 \qquad q = 1 \qquad u = 22$

$$\hat{B} = \begin{pmatrix} 1 & 0 & 1 \\ 0 & 66 & -4 \\ 5 & 80 & 0 \end{pmatrix} \qquad \hat{\delta}^* = 10$$

$$\hat{G} = \hat{B}^{-1}\hat{N} = \begin{pmatrix} -14/10 & 66/10 \\ 1/10 & -4/10 \\ 4/10 & -66/10 \end{pmatrix} \equiv \begin{pmatrix} 8/10 & 6/10 \\ 1/10 & 6/10 \\ 4/10 & 4/10 \end{pmatrix} \pmod 1.$$

$$\hat{g}^{o} = \hat{B}^{-1}\hat{b} = \begin{pmatrix} 892/10 \\ -53/10 \\ 58/10 \end{pmatrix} \equiv \begin{pmatrix} 2/10 \\ 7/10 \\ 8/10 \end{pmatrix} \pmod 1$$

$$\hat{d}_{\hat{N}} = d_{\hat{N}} - d_{\hat{B}}\hat{G} = \begin{pmatrix} \frac{1}{2} & 1 \end{pmatrix}$$

Following the shortest path problem (over a graph of 10 nodes), we get

$$x^*_{\hat{N}} = (x^*_3 \; x^*_5) = (1 \quad 1)$$

$$x^*_{\hat{B}} = (x^*_1 \; x^*_2 \; x^*_4) = \hat{C}(\hat{g}^{o} - \hat{G}x^*_{\hat{N}}) = (4 \; 5 \; 12) > 0$$

$$z^* = 17$$

2) $q_3 = \frac{1}{4}$

$t = (4 \quad 66) \qquad h = 0 \qquad q = 1 \qquad u = 21$

$$\hat{G} = \begin{pmatrix} 2/8 & 66/8 \\ 0 & -4/8 \\ -10/8 & -66/8 \end{pmatrix} \equiv \begin{pmatrix} 2/8 & 2/8 \\ 0 & 4/8 \\ 6/8 & 6/8 \end{pmatrix} \pmod 1$$

$$\hat{g}^{o} = \begin{pmatrix} 938/8 \\ -56/8 \\ -178/8 \end{pmatrix} \equiv \begin{pmatrix} 2/8 \\ 0 \\ 6/8 \end{pmatrix} \pmod 1$$

$$\hat{d}_{\hat{N}} = \left(\frac{6}{8} \quad \frac{10}{8}\right)$$

The set of the 8 graph nodes is given below, ordered with

respect to their costs and together with the correspon-ding values of x_N^* :

nodes	$\begin{pmatrix}0\\0\\0\end{pmatrix}$	$\begin{pmatrix}2/8\\4/8\\6/8\end{pmatrix}$	$\begin{pmatrix}2/8\\0\\6/8\end{pmatrix}$	$\begin{pmatrix}4/8\\0\\4/8\end{pmatrix}$	$\begin{pmatrix}4/8\\4/8\\4/8\end{pmatrix}$	$\begin{pmatrix}6/8\\4/8\\2/8\end{pmatrix}$	$\begin{pmatrix}6/8\\0\\2/8\end{pmatrix}$	$\begin{pmatrix}0\\4/8\\0\end{pmatrix}$
x_N^*	(0 0)	(0 1)	(1 0)	(2 0)	(1 1)	(2 1)	(3 0)	(3 1)
cost	0	10/8	6/8	12/8	16/8	22/8	18/8	28/8

In fig. 2 the shortest path tree and the branch and bound tree (for every node we give $x_{\hat{N}}^*$, $\bar{x}_{\hat{B}}^*$, $x_{\hat{B}}^*$, z) (see section 2.5) are shown.

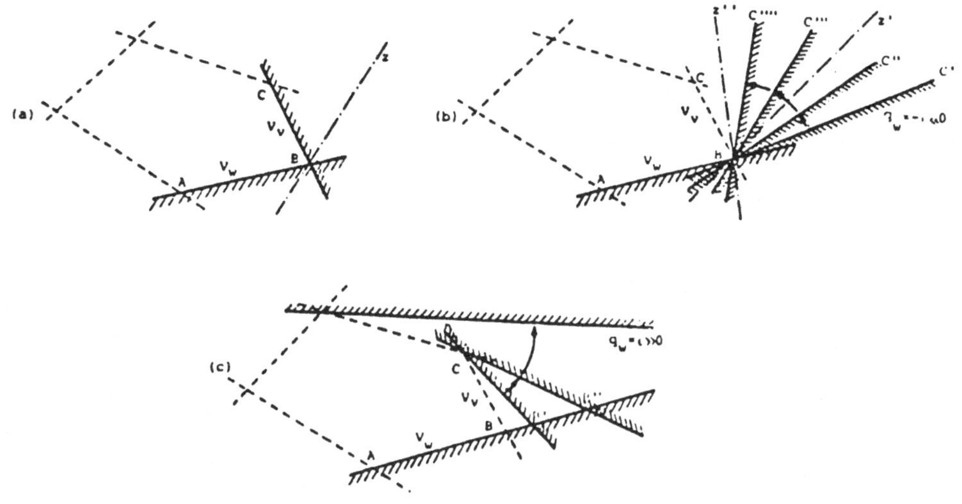

fig. 1

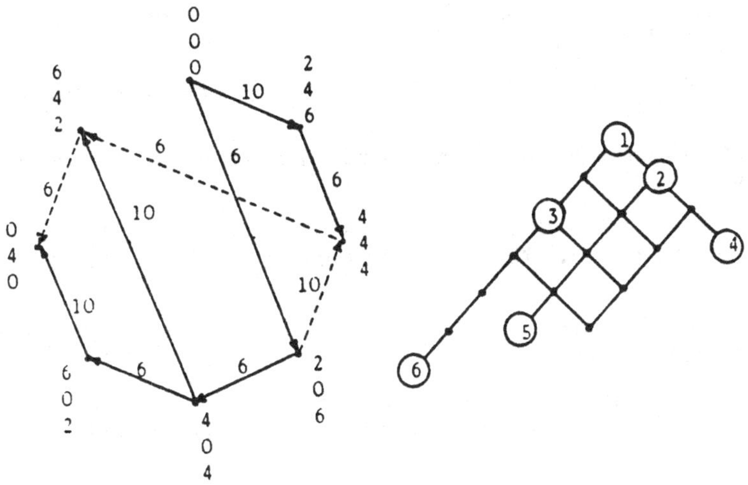

Subproblems	1	2	3	4	5	6
Objective function	22	18	19	17	15	16
$x^*_{\hat{N}}$	$\begin{pmatrix}1\\0\end{pmatrix}$	$\begin{pmatrix}3\\2\end{pmatrix}$	$\begin{pmatrix}5\\0\end{pmatrix}$	$\begin{pmatrix}1\\4\end{pmatrix}$	$\begin{pmatrix}7\\2\end{pmatrix}$	$\begin{pmatrix}9\\0\end{pmatrix}$
$\bar{x}^*_{\hat{B}}$	$\begin{pmatrix}117\\-7\\-21\end{pmatrix}$	$\begin{pmatrix}100\\-6\\-2\end{pmatrix}$	$\begin{pmatrix}116\\-7\\-16\end{pmatrix}$	$\begin{pmatrix}84\\-5\\12\end{pmatrix}$	$\begin{pmatrix}99\\-6\\3\end{pmatrix}$	$\begin{pmatrix}115\\-7\\-11\end{pmatrix}$
$x^*_{\hat{B}}$	$\begin{pmatrix}5\\7\\-21\end{pmatrix}$	$\begin{pmatrix}4\\6\\-2\end{pmatrix}$	$\begin{pmatrix}4\\7\\-16\end{pmatrix}$	$\begin{pmatrix}4\\5\\12\end{pmatrix}$	$\begin{pmatrix}3\\6\\3\end{pmatrix}$	$\begin{pmatrix}3\\7\\-11\end{pmatrix}$

fig. 2.

REFERENCES

[1] R.S.GARFINKEL, G.L.NEMHAUSER: *Integer programming*. John Wiley and Sons, 1972.

[2] S.ZIONTS: *Linear and integer programming*. Prentice-Hall Inc., 1974.

[3] T.C.HU, S.M.ROBINSON: *Mathematical programming*. Academic Press, 1973.

[4] D.E.BELL: *Improved Bounds for Integer Programs: A Supergroup Approach*. Research Memorande of IIASA, November 1973.

[5] A.M.GEOFFRION: *Lagrangian Relaxation for Integer Programming*. Mathematical Programming Study 2, 1974.

[6] M.P.WILLIAMS: *Experiments in the Formulation of Integer Programming Problems*. Mathematical Programming Study 2, 1974.

[7] E.M.L.BEALE, J.A.TOMLIN: *An integer programming approach to a class of combinatorial problems*. Mathematical Programming 3, 1972.

[8] G.T.ROSS, R.M.SOLAND: *A branch and bound algorithm for the generalized assignment problem*. Mathematical Programming 8, 1975.

[9] R.C.JENSLOW: *The principles of cutting-plane theory Graduate School of Industrial Administration*. Carnegie-Mellon University (Part I, Feb. 1974; Part II, Aug. 1975).

[10] S.GIULIANELLI, M.LUCERTINI: *A decomposition technique in integer linear programming*. 7th IFIP Conference, Nice, Sept. 1975.

[11] S.GIULIANELLI, M.LUCERTINI: *A group decomposition technique in large integer programming problems*. IFAC Symposium on Large Scale Systems, Udine, June 1976.

[12] L.A.WOLSEY: *Extensions of the group theoretic approach in integer programming*. Management Science, Sept. 1971.

[13] G.A.GORRY, J.F.SHAPIRO, L.A.WOLSEY: *Relaxation methods for pure and mixed integer programming problems*. Management Science, Jan. 1972.

[14] S.MACLANE, G.BIRKHOFF: *Algebra*. The MacMillan Company, 1967.

[15] H.M.SALKIN: *Integer programming*. Adisson Wesley, 1975.

[16] S.GIULIANELLI,M.LUCERTINI: *A method for reducing computational complexity of ILPC problems*. Ricerche di Automatica, Dec. 1977.

[17] F.NIVEN, H.Z.ZUCKERMAN: *An introduction to the theory of numbers*. Wisley, 1972.

[18] E.POLAK: *Computational methods in Optimization*. Academic Press, 1971.

Printed in the United States
By Bookmasters